COLLECTIBLE Beer Trays

With Value Guide

Gary Straub

Schiffer Publishing Ltd
77 Lower Valley Road, Atglen, PA 19310

Dedication

To Neggi and Schneider and the good times we shared.

To Nellie Winterfield. Nellie passed away in October of 1994. I hope she is smiling down from Heaven's Breweriana library as she reads this. She was a good friend and I think of her often.

Library of Congress Cataloging-in-Publication Data

Straub, Gary.
Collectible beer trays with value guide / Gary Straub.
    p.   cm.
Includes bibliographical references and index.
ISBN: 0-88740-840-0
1. Beer trays--Collectors and collecting--United States--Catalogs.
I. Title.
NK8459.B36S77   1995
338.7'66342--dc20   95-9043
    CIP

Copyright © 1995 by Gary Straub

All rights reserved. No part of this work may be reproduced or used in any forms or by any means graphic, electronic or mechanical, including photocopying or information storage and retrieval systems without written permission from the copyright holder.

Printed in China
ISBN: 0-88740-840-0

We are interested in hearing from authors with book ideas on related topics.

Published by Schiffer Publishing Ltd.
77 Lower Valley Road
Atglen, PA 19310
Please write for a free catalog.
This book may be purchased from the publisher.
Please include $2.95 postage.
Try your bookstore first.

Foreword

Author's warning:

Attempting to identify a beer tray's manufacturing origins may be hazardous to a collector's health. Prolonged exposure may cause severe headaches, eye damage and possible blindness, manic-depressive mood swings or, in extreme instances, chronic, flat-out, drooling, babbling, rocking, catatonic psychosis. The manufacturer identification process is addictive and often leads to a stronger age-dating habit.

Manufacturer/supplier information may or may not appear on a tray. On those trays which don't immediately reveal this information, a tedious inspection is required. More often than not, this intense scrutiny goes unrewarded. After 10 or 15 minutes spent in eye-straining examination, one commonly discovers oneself no closer to the goal than before. This fruitless result can occur 3, 7, 11, 12, or more times in a row. Occasionally, however, after ten minutes of staring through a magnifying glass under bright light, the discovery of a small, finely printed trademark symbol is made. The tray being examined will carry identification in some highly unlikely location in fine, carefully camouflaged print. This one successful effort brings inner satisfaction to reward the collector and makes all their previous labors worthwhile.

These sporadic successes constitute the underlying mechanism which makes tray identification so habit forming. Learning theorists refer to this as a variable ratio schedule of reinforcement. When used to shape behavior, this random schedule can keep a rat, pigeon, monkey, or tray collector repeating an occasionally reinforced act almost indefinitely. It is also, incidentally, a principle well-known and exploited by casinos, race tracks, jai-alai frontons, numbers games, etc. It is not merely coincidental that this conditioning principle is also used to keep people busy pulling slot machine levers.

This variable reinforcement schedule of operant conditioning is responsible for the many trays in this book bearing manufacturer/supplier information. It also keeps a significant number of people gainfully employed emptying cash boxes in Las Vegas, Atlantic City, and numerous Indian Reservations.

The reader is now fully aware of the inherent dangers involved in the tray game.

Tray identification signs became more apparent as the number of trays I catalogued increased. Some manufacturer-specific characteristics became evident when compared to other trays produced by the same company. I hope the information presented here will prove helpful.

Acknowledgments

This guide would not be available to tray collectors if not for the help offered by numerous people. The following individuals allowed me to photograph trays from their personal collections: Steve Bergquist, Steve Clifford, Gary & Donna Cushman, Barry Hunsberger, John Hutchison, Al Kogoy, Ron Nagelschmidt, Ken Ostrow, Geoff Pearsaol, Dick & Kathy Purvis, Bernie Tesmer, Ed Theberge, P.J. Weitzler, Nellie Winterfield, and Daryl Ziegler.

Kevin Connolly & Dick Draghi, two old friends, provided help, moral support, & companionship during photographic safaris throughout the Northeast. They transformed a potentially laborious task into a constant adventure, as good beer drinking buddies are depended on to do. Kevin and Dick performed their duties with a style, flair, and reckless abandon seldom seen in the countless bars, roadhouses, and taverns we visited while on our missions. Glenn Zemina transformed my hundreds of pages of nearly illegible handwritten scrawl into an orderly word-processed manuscript suitable for submission. I will forever be indebted to him for his tireless, uncomplaining efforts and assistance.

Nellie Winterfield, who passed away in October of 1994, taught me more about beer trays than I dreamed existed. I visited her often and we spent hundreds of hours talking tray talk together. She provided me with contact people, resources, and help whenever I was at a dead end. I could have never compiled this guide without her help. Her guidance, her knowledge accumulated during thirty years of breweriana dealing and collecting that she offered, and her friendship will never be forgotten. Daryl Ziegler also merits special mention. Daryl contributed over 100 hours of his time during the course of four three day visits to his home to photograph trays. His generous hospitality and help during our dozen 8/10 hour photo sessions will never be able to be repaid.

My brother, Roger Straub, provided moral support and regularly turned up breweriana collectibles found during his weekend meanderings from flea market to tag sale to antique shop.

I met many new people while working on this guide whom I now consider good friends. My time spent with Dick Purvis, Gary Cushman, Daryl Ziegler, Al Kogoy, Nellie, et. al. brings back many fond memories. I value and cherish our friendships. Thanks to one and all.

I would also like to thank all those individuals who wished to remain anonymous.

Contents

Chapter One Historical Background 6
    Prohibition 8
Chapter Two Tray Variations 11
Chapter Three Explanation of Tray Data 13
Chapter Four Manufacturer's Identification 17
Chapter Five Tray Manufacturing 20
    Supplier Information 20
    Companies/Brewery Advertising Supply 21
        Houses 21
        New Jersey 25
        Massachusetts 36
        Rhode Island 49
        New York 55
        New Hampshire 98
        Connecticut 100
        Pennsylvania 110
Bibliography 158
Value Guide 159

Chapter One
Historical Background

It is generally agreed upon amongst Breweriana historians that the production of lithographed tin beer trays began about 100 years ago from the time of this writing, give or take 5 years. By the early to mid-1890s innovations in manufacturing technology had developed to a degree that allowed for mass production of the lithographed tin advertising tray. By the end of the lithographed tray's first decade of production, the new century was already a few years underway. Our nation continued to evolve toward industrialization, with our manufacturing industries becoming increasingly committed to mass production technology. This national trend did not exclude the advertising industry. The favorable economic climate, in the course of the next 12 - 15 years, created a golden age of production, a boon for future collectors of Breweriana. A large and receptive brewing industry, philosophically indoctrinated as to the importance of advertising as a vehicle to expand product sales, spurred competition between the companies producing advertising goods. Numerous supply-side competitors were all vying for the same demand-side objective of being awarded a brewery's contract to fulfill an order for advertising pieces. This intensely competitive market during the early twentieth century gave todays collectors a wealth of trays. It gifted us with trays displaying a level of artistry beyond narrative description.

Non-collectors are often surprised by the beauty displayed, quite unexpectedly, through the medium of the common, and often overlooked, lithographed tin beer tray. Tray production steadily increased from ca. 1900 until the middle of the next decade. Hundreds of new designs were produced each year as manufacturers filled tray orders for more and more breweries. In addition to tray output, other imaginative advertising items - in impressive variety - were produced for brewers throughout the country. Breweriana items from this era include lithographed signs, calendars, foam scrapers, etched glasses, self-framed tin signs, reverse on glass items, corner signs, match safes, postcards, Vienna art plates, and tip trays. These items are highly prized by many of todays collectors.

The first decade of the twentieth century saw the beer tray's point of purchase advertising value become commonly recognized and firmly entrenched. The relatively exclusive club consisting of breweries utilizing the beer tray in advertising at the turn of the century saw its ranks swell as the years passed. By 1910, those breweries of any significant capacity which were not using advertising trays were probably eligible for admission into another club; that of endangered species.

HAMPDEN BREWING CO.
HAMPDEN ALE, LAGER
WILLIMANSETT, MASS.
MFG: ?
"WHO WANTS THE HANDSOME WAITER"
ROUND 13" PIE
Co. HB CO. 1934

For the purpose of age-dating beer trays, the hundred and few odd years they have kicked around can be further defined and narrowed. This time span is conveniently divided into periods by a number of historical events. These historical periods allow trays to be classified into distinct groups.

The most obvious event to use in categorizing beer trays, and also the broadest, is Prohibition. Trays produced prior to its enforcement in 1920 are commonly identified as "Pre-Pro".

Those made after Prohibition's repeal, from 1933 on, are referred to as "Post-Pro". These two distinctly separate groups can be further divided into more precise categories.

The notation "Pre-Pro" is frequently used to identify a tray's age/date range in the data accompanying photo examples included in this guide. As used in reference to beer trays, a "Pre-Pro" notation would correctly encompass the nearly 30 year span between the early 1890s and 1920. With the exception of porcelain examples, the trays included in the guide labeled "Pre-Pro", in an overwhelming majority of cases, were produced between 1900 and 1917. Only a handful of lithographed trays bearing age/date estimations which predate the turn of the century by more than a year or two will be found amongst this guides several hundred examples.

Breweries using lithographed trays to advertise their products during the 1890s were relatively few in number. Those that were utilizing trays generally contracted their production in much smaller quantities. Only a limited number somehow survived the trials and tribulations of time to become cherished and prominently displayed by todays collectors. They are generally quite rare.

Of the survivors, still fewer exist that fit the cliched descriptions "having aged gracefully", "improved with age like fine wine", or as having acquired "a distinguished look". Many display the beer trays equivalent symptoms characteristic of old age. These include any one, or combination, of the following: chipping, scratching, peeling, fading, rusting, spotting, pitting, crazing, denting, scuffing, hazing, or staining.

Many early beer trays also exhibit scars inflicted by man. These disfigurations resulted from various experimental attempts to alter the tray's intended purpose. Ill-advised attempts were made to modify them for use as targets, paint brush holders, dust pans, small change banks, planters for seed germination, and as sieves or strainers. I feel it fairly safe to speculate that the inspirational muse compelling Rod Stewart to record the song "You Wear it Well" was not the superb condition of a century-old beer tray that he was admiring.

Circumstances affecting the manufacture of beer trays, culminating in national Prohibition, began growing in influence with each passing year of the twentieth century's second decade. The winds of change had begun to shift unfavorably for the beer tray long before the actual enforcement of Prohibition began in 1920. The 18th Amendment was actually passed in December of 1917. It took the following 13 months, until January of 1919, to gain enough state ratification to become law. Any last hope of reprieve died in October of the same year when Congress overrode President Wilson's veto. Three months later, in January of 1920, a year after its ratification, national enforcement began.

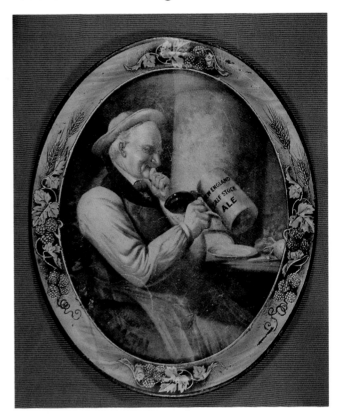

NEW ENGLAND HALF STOCK ALE
HARTFORD, CT.
THE H. D. BEACH CO.
OVAL 14" x 17"
1905 - 1915

Lesser known historical events preceded those already related, which set the stage for their eventual unfolding. Initially a phenomenon most common to our southeastern and northern New England states, a number of Prohibition laws on an individual, state-legislated level were passed well before national ratification occurred. By 1916, statewide Prohibition laws had been passed by twenty-four (exactly half) of our states. The course of international events made our eventual participation in WWI increasingly apparent. Our 1917 declaration of war caused a widespread and growing anti-German sentiment in America. The ancestry and Teutonic heritage of many of our nation's brewers made their patriotism suspect and subjected them to general feelings of prejudice.

The advertising advantages to be gained by linking a brewer's product with a German surname were questionable at best, and thought likely to have counterproductive results by many brewers.

The combined effects of a legislatively reduced market base, the brewing industry's uncertain future, and this nation's widespread anti-German sentiments took a heavy toll on beer tray production. This plummet in tray manufacturing is evidenced by the scarcity of beer trays bearing copyright dates between 1917 and 1920. The economic conditions that fostered the beer tray's rapid ascent in popularity during the century's first decade had reversed direction and reduced production even more quickly.

PROVIDENCE BREWING CO.
BOHEMIAN BEER AND CANADA MALT ALE
PROVIDENCE, R.I.
KAUFMANN AND STRAUSS
ROUND 13" PIE
PRE - PRO

In summary, the 1900 - 1917 era was responsible for a huge output of breweriana collectibles. All but a handful of lithographed trays included within, which are identified as "Pre-Pro", were manufactured during this time.

The reader should understand that this "first golden age of production" is defined in broad and general terms. A good number of lithographed trays were produced both in the 1890s and in the time after 1917 leading up to National Prohibition. The intent is to make the collector aware that trays manufactured outside this era represent a vastly reduced number of breweries. These breweries usually ordered trays in relatively small quantities from manufacturers as well.

Of total Pre-Pro lithographed tray production, I would guess 90%, and likely closer to 95%, were created during the "first golden age" years of production, 1900 - 1917.

Prohibition

It is difficult to debate the Pro-side of our 18th Constitutional Amendment and laud hosanna's upon it in praise of the many positive effects it bestowed. The "Volstead Act", named after the Congressman sponsoring it, is the name used interchangeably with any reference to the amendment. Actually, Representative Volstead was responsible for a bill defining Prohibition's legal limits and enforcement procedures. As a reward, his constituency voted him out of office as soon as election time rolled around again.

In an unforeseen, and completely unintentional way, today's breweriana collector became a major beneficiary of Prohibition. Its thirteen-year duration provides an unequivocal dividing line for use as a tool in classifying breweriana by age/date means. This classification device is admittedly useful to the breweriana collector of today. It's certainly a good deal better than anything gained from Prohibition by the rest of our nation's population. However, when its value is weighed against the thirteen year period of lost breweriana production, the price paid in return seems a bit high. Those brewers managing to survive during Prohibition did so by modifying their facilities for the production of other goods. The most common alternatives included near beer, soft drinks, block ice, malt syrups and extracts, ice cream, and a wide variety of foods. Trays advertising these new products were made for some of these former breweries (or breweries in waiting), depending upon one's frame of reference. A few examples of Prohibition era trays are pictured later. These trays include the "Congo" tray (near beer) from Haberle-Crystal and the rectangular Gerhard Lang tray advertising soft drinks.

Prohibition, from its origin as an abstract postulate to its applied practice in reality, was an exercise doomed to failure. As a concept, it is intrinsically flawed by its incompatibility with the laws governing human nature. Its usefulness to brewerianists aside, Prohibition's sole positive feature, that of serving as an example, has been too often ignored or forgotten by the legislators who have followed. Its legacy should forever exist as a testimonial, warning of the fate awaiting any further attempts to legislate morality.

In their successful crusade to deliver our nation from the evils of intoxication, Volstead's valiant vanguard of virtuous visionaries bestowed upon us additional, secondary rewards. The American populace, throughout the 1920s, witnessed long term increases in many facets defining their way of life. Several social standards skyrocketed to heights undreamed of, even by those farsighted prophets responsible. Although sobriety was not amongst the elevated, Prohibition increased organized crime, prostitution, drunk driving arrests, deaths attributed to alcohol, government employees slain in the line of duty, gin in the average American bathtub, and—in all likelihood—the number of olives garnishing that average American bathtub.

In an attempt to present an "objective and unbiased account," it is necessary to inform the reader that some standards were also lowered by Prohibition. These lowered standards included a lower average drinking age, a loss of 1 billion dollars per year in federal tax revenue, and a diminished respect for other laws. Prohibition also lowered by an estimated 60,000,000 gallons the industrial alcohol available for its intended purpose from 1924 until 1928. This industrial alcohol was being diverted for use in making synthetic liquor. Life expectancy was drastically reduced for the estimated 15,000 criminals slain for flaunting Prohibition laws. I suspect our 18th Amendment lowered the probability of bathing without sustaining injury (by slipping on olives) as well.

QUANDT BREWING CO.
QUANDT LAGER - ALES
TROY, N.Y.
CHAS. W. SHONK
ROUND 12"
PRE - PRO

After Prohibition ended in 1933 with the repeal of the 18th Amendment, the use of the beer advertising tray as a sales device underwent a resurgence in popularity.

It took most brewers several months to modify their production lines back to their original purpose. By mid-1934, over 700 breweries were back in business, and the beer tray that advertised their products became plentiful once again.

Tray manufacturers that were forced to close during Prohibition were replaced by over a dozen new companies that supplied tray demands for the decade and a half following repeal. Tray demand remained strong during the remainder of the 1930s and into the early 1940s. Then, once again, factors beyond the control of the brewing industry occurred and eventually dealt a crippling blow.

Several factors combined to negatively affect tray production, starting in the early 1940s. The most damaging of these, in no special order, were:

1. A steadily decreasing number of breweries.
2. A shortage of tin for non-military uses during World War II.
3. Closer regulation of the alcohol industry by the national government. This included passage of restrictive legislation which limited the distribution of point of purchase advertising items.
4. The electronic media—first radio and later television—which redefined the methods previously used by the advertising industry.

Each of these four factors was seriously damaging in its own right. Their combined impact, occurring in less than a decade's time, caused permanent damage to the advertising tray industry.

Tray production began to decline during WWII. This was not caused by the elimination of brewery product, as was earlier dictated by Prohibition. In this instance, the tray decline was due to the tin shortage created by the war. Metal to be used for non-military purposes was no longer as plentiful or readily available as it had been previously. Obviously, metal used for vehicles to transport G.I.'s to strategic military positions took priority over metal used to transport mugs of beer to civilian dining table positions.

Although the circumstances limiting tray output differed from those of Prohibition, the results were similar. Tray manufacturers insufficiently diversified in product, or not financially stable enough to ride out the hard times, fell by the wayside. Unlike the 1933 resurgence after the repeal of Prohibition, however, renewed post-war tin availability enticed no significant number of replacements to fill the decimated ranks of the "Tray Team" manufacturing companies.

This lack of manufacturing opportunists eager to grab a newly available market share of the freshly thinned tray industry was not produced by an absence of foresight on their part. I suspect the lack of tray manufacturing newcomers resulted from a steadily eroding market base. Of the 750 +/- breweries operating in 1934, only about 590 lasted until 1940. By 1950, 200 more were no longer in business.

With no sign to indicate any reversal of this downward spiral, the path toward future long term prosperity and security for tray producing newcomers did not exactly appear paved of gold and lined with roses. Now equipped with the gift of 20/20 hindsight, one can appreciate their perceptive intuition in wisely deciding not to sink into lifetime debt in order to purchase or begin a "can't-miss" tray manufacturing gold mine. Early in the 1970s, only around 50 brewing companies remained in business.

The winds of change began to shift again shortly before the 1980s began. Small microbrewery Davids began challenging the national mega-brewery Goliaths. These microbreweries succeeded in finding a demand for their products among the American beer drinking public. In less than a decade, their numbers grew and reversed the direction that the brewing industry had followed since 1933. Microbrewers also were a Godsend for many small regional independent breweries that were struggling to survive. Microbrewers often subcontract the actual brewing and packaging of their products to independents, providing them with an additional source of income. This has rescued some struggling independents who otherwise would have already been listed in the brewery obituary column. In 1980, it seemed only a matter of time before the small independent brewer was a thing of the past. Unable to compete against the brewing giants with their endless advertising funds, many suffered steadily decreasing sales. In 1980 I would have predicted that by the year 2000 the mega brewery advertising sledgehammer would result in there being only a dozen or so breweries remaining in this country.

Following the lead of their microbrewery allies, many independents began offereing their own high quality brews to the public. Contract brewing and new product profits made it possible to survive despite the best efforts of the nationals to increase their market share at the expense of the small independents. The nationals' attempted "genocide" of the small brewers ultimately failed. Today, many have rebounded quite successfully and are in good health. I suspect upstate New York beer drinkers scanning the brewery obits in a Rochester paper will not soon see their local brewery has fallen victim to what I would call "Geneseeocide."

The successful efforts of early microbrewers did not go unnoticed. They inspired a still-growing grass roots groundswell of new generation brewmasters who share a common belief in a revolutionary concept. They believe that, if offered a choice, some beer drinkers prefer a hand-brewed porter, stout, weiss, dopplebock, wheat, dark ale, etc. over the countless homogeneous, mass-produced beer clone sextets offered by the big boys

NARRAGANSETT BREWING CO
NARRAGANSETT LAGER AND ALE
PROVIDENCE, R.I.
MFG: ?
ARTIST: DR. SUESS
ROUND 12"
1940s

UNION BREWERY, HOWELL & KING CO.
PURE BEER, ALE AND PORTER
PITTSTON, PA.
THE MEEK CO.
"GOOD FRIENDS"
RECT 13" x 18"
1901 - 1909

Chapter Two
Tray Variations

This guide includes color photos and accompanying descriptive data for over 600 Northeastern U.S. beer trays. Around 95% of these trays are examples of the type produced using the manufacturing process of lithography. Lithography employed presses designed to transfer a paint surface coat onto the metal or tin base material composing the tray. This process provided a vehicle for brewers to advertise their products which was directly focused on their potential target market. The beer tray's multi-colored advertising design gave brewers a high-visibility tool to increase brand name recognition and hopefully influence customers to try their products.

The beer tray is an item designed to be used in sales locations. Its advertising relatives include tap knobs, coasters, neons, foam scrapers, napkins, menu sheets, openers, tip trays, and register clocks amongst many others. Collectively, this group of items is known as point of purchase advertising. As opposed to billboards, magazine and newspaper ads, or radio and TV messages, point of purchase items were distributed to taverns, bars, and retail outlets to influence people at those locations to purchase alcoholic beverages. Some point of purchase items, such as mirrors, signs, and statues were distributed simply as decorative pieces to increase brand name familiarity. A large percentage, however, served the dual purposes of advertising and performing a practical function. They were distributed to proprietors to be used in various ways while serving their establishment's clientele. If those old trays you collectors proudly display on your Breweriana room walls could only talk, I would imagine we would hear some pretty interesting war stories, recollected from the old days when they were on active duty in the service of some ancient, long-closed bar or tavern. Collectors would learn the circumstances behind each battle scar they display, suffered while transporting replacements drawn up from the draft lines to the front whenever ordered. Collectors would hear tales of the tray's responsibility to also carry drained veterans, old mugs, and dead soldiers back to the rear.

The lithograph process has undergone many changes over a century's passing. The original limestone plates used in printing were later replaced by zinc plates, which in turn were rendered obsolete by the silk-screen process. The manufacturing process became ever faster, more automated, and less dependent on the various tradesmen whose skills were originally required to produce the lithographed beer tray.

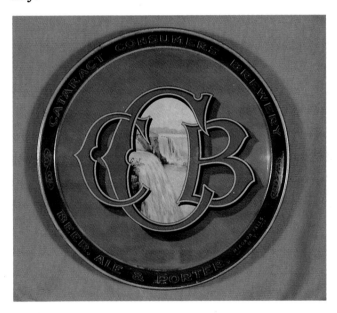

CATARACT CONSUMERS BREWERY
BEER, ALE & PORTER
NIAGARA FALLS, N.Y.
CHAS. W. SHONK LITHO
ROUND 12"
1905 - 1913

This guide also contains examples of a few dozen trays which date back from the turn of the twentieth century, +/- a decade, created by a different means. These trays, commonly called "Porcelain," were manufactured by baking an enamel finish over a metal base. There are also a few examples of "Bakelite" trays dating from the WWII years when tin was hard to come by.

I have included just one example of another type of tray, those made of brass or those having a silver-type plated finish. These trays also date from the late nineteenth and early twentieth centuries. From an advertising viewpoint, their effectiveness was inferior to the others in terms of the visibility of their message. These trays, not having the advantage of contrasting colors, utilized etching or engraving techniques which were stamped into the tray surfaces to communicate their messages.

Porcelain, brass, silver-plate, and Bakelite trays all had disadvantages which made them markedly inferior to those made using lithography. The enamel over tin tray's distribution was overwhelmingly restricted to Eastern United States and Canadian brewers. "Porcelains" were most commonly used by brewers from New York, Connecticut, Massachusetts, and Rhode Island. The largest manufacturer of these trays was the Baltimore Enamel and Novelty Company. Despite the illusion of durability implied by their heaviness, "porcelain" trays were quite fragile. If dropped or banged around, they chipped and cracked easily. They were also prone to discoloring, staining, scratching, and wearing of their enamel surfaces.

Trays made of brass and those of silver-plated tin were largely at the mercy of the angle of reflected light in order to make their advertising message visible. Financially speaking, in terms of cost-effectiveness, both porcelain and engraved metal trays were vastly more expensive per unit than their lithographed relatives.

The short-lived bakelite trays would likely never have been produced at all if not for the circumstances dictated by WWII. Despite the metal shortage during the war years, these trays never received widespread use and thankfully vanished the instant tin once again became available.

Completely unrelated to rarity, photographic limitations, or any other aforementioned reason, one other type of tray is excluded from this guide. Trays made of the latest innovation in material - plastic.

Plastics technology is running completely amok and has victimized the innocent Breweriana collector. It has spawned the plastic beer tray - a hideous, mutant aberration. Based on the number of these faux-tray androids infesting Breweriana shows, I would guess that they are multiplying at about twice the rate as your average plague of locusts.

Since it is available, I cannot resist this opportunity to express an opinion which is likely shared by a majority of collectors. Plastic beer trays possess absolutely no artistic beauty and are totally devoid of any redeeming aesthetic qualities whatsoever.

Plastic, as a synthetic material, does not go away easily. It is virtually immune to the elements. Melting a plastic tray will open an equivalent tray-sized porthole in our ozone layer. When they are dumped from garbage barges, even the ocean floor exhibits signs of being an intelligent life form. It spits them floating to the surface looking like they advertise some newly issued Lazarus lager. Lastly, plastic has a molecular half life of somewhere in the neighborhood of 500 years.

The plastic beer tray, along with Pampers, disposable plates, and about half of the total parts making up any new car will be decreasing the property values of local landfills, and future Breweriana collections, for centuries after all traces of our generation of collectors has vanished from the face of the Earth.

Given the beer tray's developmental history during the century plus of their existence, one can trace the metamorphosis from solid brass, plated metal, and porcelain, to quality lithographed pieces, to mass-produced silk-screened examples, to plastic discs so hopelessly bland and neutral that they are incapable of eliciting any kind of emotional response. Given this sequential pattern, it seems highly unlikely that the beer tray will evolve next into a tray composed of semi-precious metals with an inlaid Mother of Pearl advertising message. Although I am admittedly not blessed with an ordered, logical mind equipped to intuitively grasp any kind of evolutionary progression theory, I am inclined to lean toward the beer tray of the future being "improved" by:

1. Being constructed out of compressed, high-density Styrofoam.

or

2. Evolving into convenient, disposable models made of pressure treated or corrugated cardboard coated with a beer-resistant waxy finish.

Thus, it becomes the responsibility of today's collectors to preserve and safeguard the beauty and artistry of a bygone era. It is important to pass along to future generations of collectors the history and examples which were produced during an earlier time and exist today as a testimony to an art form created by the American Advertising Industry.

The trays contained within are pieces of a puzzle which, when put into place, reveal an overview of an industry. They trace its development from local institution to the present day, multi-branched, national, mega-corporations.

Chapter Three
Explanation of Tray Data

All trays included in this guide have identifying data accompanying their photographs. This information is often too small or too close in color to a tray's background to identify from its photograph.

Tray data is given in the following order:

1. Brewery - The full name printed on the tray is given. The exact name of the brewery is important for purposes of age dating. Donald Bull's *American Breweries* was the reference used to determine a brewery's dates of production. Seemingly trivial differences sometimes provide important information. For example, from 1889 - 1920, in Newark, New Jersey, Hensler's Brewery conducted business as "Joseph Hensler Brewing Co.". Upon resuming business after Prohibition ended in 1933, the brewery name was identified as "*The* Joseph Hensler Brewing Co.".

2. Brand name - Brands are listed whenever given. Some trays do not identify specific brand names, but instead promote the name of the brewery. Others identify their product line (lager, ale, porter, dark, pilsner, bock, etc.). This line provides whatever information appears on the tray itself. Brand name information is commonly printed on the trays inner (and sometimes outer) sides. In many examples, these brand names may not be visible because of the angle from which the trays were photographed.

3. Location - The town or city, and state location of the brewery the tray advertises. The location of the original brewery is listed first in cases of multiple addresses. (Example: Liebmann Breweries, Inc., Brooklyn, N.Y., Orange, N.J.)

4. Tray Manufacturer/Pertinent Information - If the tray manufacturing company is printed, this information is provided. Unfortunately, many trays bear no logo, trademark symbol, or numeral code to positively identify their manufacturer. Over 50 company identities are listed on the trays contained in this book. This listing identifies the name of the company accredited on the tray. A number of companies listed were not involved in the tray's actual manufacturing process. They were advertising supply middlemen that breweries contracted to provide promotional products. They, in turn, subcontracted manufacturers to produce these goods. There is a good deal of evidence to suggest that trays bearing the N.Y. Importing Co. name were actually manufactured by the Burdick Company, for example.

The identity of a tray's manufacturer or supply company may provide additional information helpful in establishing an age-date for it. The economic ruin wrought by

Prohibition on the alcohol industry produced ripple effects. Shock waves extended outward to cause financial hardship for secondary, alcohol-related businesses and industries as well. Advertising trays ordered by brewers and distillers sometimes represented such a vital percentage of a tray manufacturer's total output that they did not survive without these accounts. The financial fortunes of alcohol-related industries during the era of Prohibition paralleled those of brewers and distillers. After Prohibition ended in 1933, and the dust finally settled, the toll suffered by tray manufacturers could be accurately accessed. The final tally revealed that some tray manufacturers/suppliers survived and that new companies quickly filled the positions formerly occupied by businesses lost to Prohibition.

Included is a list identifying all tray manufacturers/supply companies located in the United States that I am aware of. These are listed alphabetically and, in some cases, provide addresses or years of business operation. This alphabetical list is followed by a breakdown of these companies into 3 groups: those operating only before Prohibition, those operating exclusively post-Prohibition, and those which are identified on both pre- and post-Prohibition trays. In some instances, usually involving companies accredited on a very small number of trays, I could not locate sufficient information to allow me to categorize them.

In addition to manufacturer/supplier identification, information usually not visible on the tray's photo example may also be included. This data may identify a tray's stock number, artist, title ("Roderick", "Griselda", "The Invitation", "Beauty and the Beast") or any other distinguishing features it exhibits.

5. Size/Style - Tray measurements (from outer rim to outer rim across the middle), vary considerably between manufacturers. Round trays listed in this guide as 12" or 13" are often 1/4" or 1/2" larger or smaller. Oval trays most commonly measured either 13" x 16" or 14" x 17". Rectangular trays, in an overwhelming majority of cases, were produced in an 11" x 14" (actually 10 3/4" x 13 1/2") size. Tray dimensions given in this guide are not exact. They define distinct categories recognized and used by collectors to differentiate general size groupings. A round tray identified as 12", although possibly measuring 12 1/2", is distinctly recognizable and not easily confused with an example listed as being 13". I would guess that probably 96 - 98% of all round trays fall into either the 12" or 13" categories of size groupings. A handful of 14" round trays, most produced by the Novelty Advertising Company (Coshoncton, Ohio) also appear in this guide. Over the years, a small number of trays were manufactured in various off-sizes, none of which ever achieved widespread popularity or usage. Examples include a Fehr's 16" round pie made by H.D. Beach, a Robin Hood 8" x 11 1/2" oval produced by Electro-Chemical Engraving, and an Oertels 13" x 18" rectangle, also from H.D. Beach.

The second distinction identified here, along with size, pertains to the style in which a tray's sides are shaped. This information applies only to the round shaped trays presented in this volume. Following their size measurement specification, some round trays are listed as "Pies". This term is commonly used in a descriptive context when referring to trays having curved or arced sides. "Pie" is used in the broad sense here, making no distinction between the "standard" and "reverse or inverted" varieties. Trays not specifically identified as pies make up the other category; those characterized by having a straight-sided construction style. The classification "straight-sided" shouldn't be misinterpreted as meaning vertically sided. Usually, their sides are slanted outward at various angles. Neither does the term apply to upper rim construction. All trays have flat or curved rims, the edges of which curl out and then back around underneath the upper rim surface.

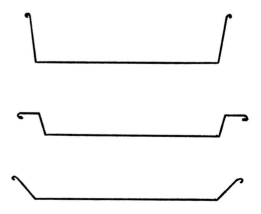

All three of these examples depict trays with straight sidewall construction. The first two (one deep and the other shallow) have much more sharply angled sides than the third. Most often the sides of straight-sided trays are angled outward at a 75-80 degree pitch.

The "Pie" styled tray was widely used prior to Prohibition and remained very common throughout the 1930s. The limited availability of tin for non-military use during World War II created an overall decline in tray production during these years. For the pie styled tray, however, it seemed the war never ended.

Pie-styled tray examples continued declining in number through the late 1940s. By the mid-1950s, straight-sided trays were being used almost exclusively. Over the last 40 years—and continuing today—round, straight-sided trays have been the only shape and style manufactured; these relatively recent trays have been limited to variations in size alone, measuring either 12" or 13" across. There are very few exceptions—other than the reproductions of earlier trays which are not easily confused with the originals.

"Pie" styled trays. In the vast majority of cases, the following statements apply: round porcelain trays from U.S. breweries are "reverse or inverted pies"; inverted pies are usually shallow, measuring only 1/2" to 5/8" from the upper rim to the base; the inverted pie tray was most popular prior to prohibition. Post-Pro trays that are pies are of the "standard pie" design with concave sidewalls, usually measuring 1 1/4" deep, as are the 12" Post-Pro straight-sided trays. Post-Pro straight-sided trays measurin 13" in diameter are 1 1/2" to 1 5/8" deep as a rule.

6. Age/Date Estimate - This category of information provides either a specific year of copyright which appears on a tray or gives a date-range estimate identifying the time period during which a tray was most likely manufactured. Some tray manufacturers/suppliers included a copyright date in their identification logo. Trays bearing their date of production, however, prove to be the exception rather than the rule. As a ballpark figure, I would guess that only about 15% of all trays carry specific date information. The tray manufacturers that most frequently provided dates of production or copyright years were American Art Works, Electro-Chemical Engraving, and Passaic Metal Ware. A number of Meek Co. trays, as well as some from Mayer and Lavenson, also bear copyright dates.

Manufacturers which rarely or never printed age-related information on their trays included Beach, Shonk, Haeusermann, and Burdick. Whenever possible, age-range periods were narrowed in scope by cross referencing the operational years of a tray's producer and the production dates of the advertised brewery.

Example: A 14" x 17" oval tray is identified in large print on its top "pie-style" rim with the following:

"Kostenbader Export Beer"

Centered on the lower rim is the information:

"Catasauqua, Pa."

Closer inspection reveals further information: Centered on the lower tray face is the number 42, the identifying stock number. (This tray was also used by Graupner Brewery as well as by a number of others). A few inches to the right of the stock number is the Meek trademark symbol. The Meek Company manufactured trays from 1901 - 1909, before being renamed American Art Works. From the stock number alone (and many of the Meek or later American Art Work products lacked the company name or logo), we can identify this tray as either a Meek or American Art Works product. In this case, however, the information identifying the company is provided, thus making further investigations unnecessary.

Donald Bull's *American Breweries* provides the following information about the brewery advertised on this tray:

Schaefer & Kostenbader, Eagle Brewery
1867 - 1872
Kreutzer & Kostenbader, Eagle Brewery
1872 - 1876
Herman Kostenbader, Eagle Brewery
1876 - 1902
*Herman Kostenbader & Sons, Eagle Brewery
1902 - 1920
Eagle Brewing Co
1934 - 1964

The year 1901 was also the final year Herman Kostenbader operated before being joined by his two sons. The tray was likely the material result of his sons' enthusiasm and their "new-fangled" methods to increase business. The operating years of this manifestation of the brewery began in 1902 and lasted until Prohibition forced Herman Kostenbader & Sons, Eagle Brewery out of business in 1920. Since the Meek Co. logo was included on the tray as well, and that logo was in use only prior to the company's rechristening as American Art Works in 1909, this tray, in all likelihood, was made sometime between 1902 and 1909.

There are a few instances when an exact date is given for a tray as a result of lousy luck, lousy management, lousy beer, or lousy something, which resulted in a breweries first and last bunged barrels of brew both having a common year of production.

Example: Molter's What Cheer Brewery in Providence, R.I., brewed "Princess Ale" and "What Cheer Lager" between the years 1910 and 1910. I suspect "Princess Ale" was probably shortened from "You Didn't Just Clean Your Beer Ta *Pr In Cess* Pool P*ales* Again, Did You?". "What Cheer Lager" simply eliminated its previous punctuation mark from the former "What Cheer?, Lager", I would guess.

In some cases, when not included in a tray's manufacturing logo, age/date information can be found cleverly concealed in a variety of unlikely locations. During my research it has been variously discovered in brewery information, with signatures of artists, and as a detail of an overall theme (on a license plate, calendar, sheet music scroll, bottle label, etc.).

A bright, even light and a jeweler's lens or good magnifying glass are required equipment for any aspiring tray sleuths possessing less than 20-20 bird of prey vision capability.

Chapter Four
Manufacturer's Identification

A time line graph tracing the progression of a few manufacturers/suppliers may provide some historical perspective:

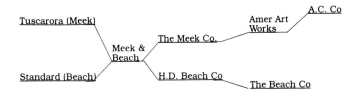

The Meek Company sometimes is accredited using its full company name on trays. In other examples, this trademark symbol is used:

American Art Works used its full name, or, in some instances the A.A.W. initials. Other A.A.W. trays bear their Trademark symbol:

I am not aware of any trays from A.A.W.'s later Newark subsidiary, American Colortype Company, that used any identification other than "A.C. Co."

The Meek Company had an inventory of many stock pictures that breweries ordering trays could choose to advertise their product. Smaller brewers often used stock designs to spare the expense of commissioning an original artistic creation and the production costs of color breakdown and transference of the artist's work onto lithograph plates.

Stock trays normally carry an identification number, usually located at six o'clock on the tray's face. One popular and widely used example of a stock tray portrays a 12 point buck pictured with an evergreen forest at sunset in the background. This tray was produced by the Meek Company and later bore the American Art Works identification after the company was renamed. This particular tray design is identified as stock #77. This majestic stag, #77, was lithographed on numerous trays, advertising brewers including Hand, Elk, South Bethlehem, and Kostenbader. If this beer deer actually drank all the different products he advertised, this never in a rut, staggered stag would have had a nose like Rudolph's. Number 77 also advertised White Bock Beer on a tray used by Jacob Hornung. He was their Stock Bock Buck.

Stock numbers occasionally are seen on Shonk and Beach Trays, but in the vast majority of cases, trays identified solely in this way were produced by Meek or American Art Works. The use of stock numbers usually occurred on trays made prior to Prohibition and the practice did not often carry over, if at all, to Post-Pro trays. The demise of the A.C. Co. disproves the theory that the Meek shall eventually inherit the earth, or even the beer tray industry, in this case.

The Burdick Co., which produced a significant number of trays during the 1930s and into the 40s, often used the following trademark symbol in lieu of their fully printed company name as identification:

Burdick Trays, to my knowledge, are all either 12" round straight-sided or 13" round pie-styled examples.

The only exception is the style variation used for the factory-scene tray this company manufactured for Stegmaier. This tray is a lightweight, thin-gauge tray with very shallow sides. At the top the sides extend horizontally to form a flat inch-wide rim. This seldom used tray design is identical in dimension, shape, and thin-gauge construction to the Joyce's Perfection tray bearing the N.Y. Importing Company name. N.Y. Importing Company was an advertising supply house that most likely sub-contracted Burdick for manufacturing purposes to fulfill brewery tray orders. Trays bearing the N.Y. Importing

Company name are most commonly 13" round pies sharing similar construction characteristics with those credited to Burdick.

Chas. W. Shonk Company Litho produced a large number of Pre-Prohibition trays. A clue to the Shonk Company's disappearance in Post-Prohibition tray manufacturing appears on just one tray I am familiar with. During Prohibition, the Haberle-Crystal Brewing Company, of Syracuse, survived by brewing a near beer called "Congo," in addition to other products. A Congo advertising tray from the Prohibition era identifies its manufacturer as "Shonk Works-Amer Can Co. - Haywood, IL." Although unable to discover any substantiating evidence, it seems reasonable to conclude that Shonk was absorbed by American Can at some point during Prohibition.

American Can Company sometimes centers their trademark on a tray's back side. Another location frequently used is the "6 o'clock" lower center of a tray's inner sidewall base. Earlier trays made by this company were usually labeled "Am. Can Co." or "Amer. Can Co.". They later used this familiar mark:

In some instances, trays from American Can simply carry a coded identification, the most frequently seen is "71 A". This refers to the location of the branch division which actually manufactured the tray.

Those trays made by the Continental Can Company that bear an identifying credit almost always use the Company's trademark of three C's, each one slightly smaller and nested within the largest C.

Since the early 1950s, either American or Continental Can have been the manufacturers responsible for almost all the lithograph on tin beer trays produced.

Baltimore Enamel and Novelty Company manufactured the vast majority of enamel over metal, or "porcelain" trays. These trays were used almost exclusively by brewers located in the Northeast and Mid-Atlantic states, as well as by many Canadian breweries. The "porcelain" tray seems to have been quite popular with Canadian Brewers, based on the many different examples that I am aware of. "Porcelain", a term not exactly accurate, is the name commonly used by tray collectors in reference to enamel over tin style trays. Trays having this compostion material are, to my knowledge, either round "Pie" or oval shaped. In most cases, Baltimore Enamel's identification appears centered on a trays back surface.

Electro-Chemical Engraving Company, Inc., a Post-Pro manufacturer, regularly printed their company credits across the bottom of a tray's face side. Their name was arced along the rounding tray face at the base of its sidewall. Following the Company name, Electro-Chem. often included a production date on their trays.

Example:

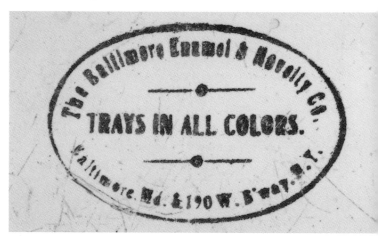

The 12" round, cream colored background/red scripted "Stanton" tray (Troy, N.Y.), bears the following I.D.:

Electro-Chemical Engraving Co., Inc. New York, N.Y. 4-36

Although not all of this manufacturer's trays include date information, a significant number do. Listing specific numerical month/year date information is an Electro-Chem trait which is unique amongst tray manufacturers.

For some reason which has completely baffled me and, I concede, may well be beyond my ability to intellectually grasp at all, some tray manufacturers concealed their credit identifications very carefully. Their imaginative camouflage techniques have grudgingly won my complete respect and admiration. Colorations and adaptations designed by Mother Nature to disguise many species of her animal kingdom pale in comparison. To enable the reader to fully understand this phenomenon, let me present a common experience, analogous in principle.

After eventually being coerced into repainting a room (or more), the clerk at the paint store responds to your request for a gallon of off-white paint by providing you with a color chart. The few hundred examples the chart includes would all be called "off-white" by the average American citizen. The difference between any one example and the 3 or 4 on either side of it that are closest in tint or shade is invisible to the naked eye. Dirty Dishwater Off-White appears identical to B.V.D. Underwear Washed With Black Socks Off-White.

When this example is applied to beer trays, it produces the following results:

When a tray's background is colored a light Sky Blue shade, the manufacturer often decided to print its identification credits in Robin's Egg Blue, Caribbean Sea Blue, Paul Newman's Eyes Blue, "Its A Boy" Cigar Band Blue or '57 Chevy Aquamarine Blue. In comparison, a spotted fawn, stick caterpillar, chameleon, or nesting woodcock is about as camouflaged as a frisbee would be if displayed amongst 20 beer trays at a Breweriana Show.

Clues to manufacturing company identification may be provided by construction characteristics common to trays made by that specific company. Distinctive production features, such as paint thickness, rim edge shape, sidewall height, width gauge of tin used, sidewall construction variations of scoring, and smoothness, upper rim shape, etc. can all offer evidence in efforts to determine a trays maker.

The comparison of several trays will reveal individualizing differences in design between manufacturing companies. If a tray is turned upside-down, one can easily discern differences in how a tray's rim edge is finished. Some tray makers produced their trays with rim edges curled up under the rim, some simply flattened out and faced inward. Early Beach trays had a wide, flat upper rim characterizing their design. To my knowledge, the square tray, measuring roughly 14 x 14, unlike its round, rectangular, ing roughly 14 x 14, unlike its round, rectangular, or oval relatives, was a shape option offered exclusively by one manufacturer, The Meek Co., or later its renamed manifestation, American Art Works. One or the other is accredited on all the square shaped trays made in the U.S. that I am familiar with.

Chapter Five
Tray Manufacturing

Supplier Information

The following chart is based on the research done for the Northeastern trays included in this guide. A number of the manufacturers' supply houses listed are credited on numerous trays. Obviously, two or three trays are not a sufficient sample number to base an ironclad judgement on. A distinct possibility exists that trays unfamiliar to me were produced which would alter their manufacturers group placement in this chart. Paradoxically, in some cases, the limited number of examples accredited to some manufacturers provide the grounds upon which I based my judgement. If a company identification appears on just a handful of trays, comprising only a fraction of 1% of the 600+ total sample group included, it is probable that:

1. There will not be several dozen other trays from this company amongst trays not included in this book.

2. The maker was likely a small local manufacturer that was contracted by a nearby brewer to provide a relatively limited number of advertising trays. I would be very surprised to find evidence indicating that the phones were ringing off the walls at Pilgrim Novelty, Grammes, Nordham, Fenstermaker, et.al. with tray orders pouring in from brewers across the nation.

I compiled this chart with particular care, and, having a somewhat vested interest, sincerely hope it is largely accurate. It was not particularly difficult to determine that the consequences of error were best to be avoided. Proof of my ignorance would likely produce skepticism and doubts regarding the overall validity of this reference. Ultimately, I would surely be denied the secondary benefits my newly perceived status as a tray authority respectfully offers. So, based upon my available research information, I present the following production era breakdown of Tray Manufacturers/Suppliers.

Companies/Brewery Advertising Supply

Tray Manufacturers/Suppliers Operating Exclusively Pre-Pro

American Art Sign	Ivan Nordham
Bachrach	Passaic Metal Ware
Baltimore Enamel	Pilgrim
Haeusermann	Roesch
International Adv.	Sentenne and Green
F.E. Marsland	Chas. W. Shonk.
Mayer and Lavinson	Standard Adv.
Meek	Tuscarora
Meek and Beach	Wolff and Lawrence

Tray Manufacturers/Suppliers Operating Exclusively Post-Pro

A.C Co.	Mason Can Co.
ACME	Metallograph

Burdick
CCC
Electro-Chemical
Fenstermaker
General Brewers Supply
N.Y. Importing Co.
Offerman
Owens-Illinois
Tindeco
Universal

Companies Spanning Prohibition

American Art Works
Beach/American Can Co.
Novelty Adv. Co.
Kaufman and Strauss and Niagra, may also be included in this group.

    I would appreciate hearing from collectors who know of additional tray manufacturers. Also, I would welcome information about the preceding companies and any trays produced by them which could alter their production era grouping.

Houses

American Art Sign Co. - Brooklyn, N.Y.

American Art Works, Inc. 1909 - 1950
Coshoncton, Ohio - later became a division of Rapid American Corp.

American Can Co. - 1901 - Present
Multiple branch locations

A.C Co. (American Colortype Co.) 1930 - 1950
Later became a division of Rapid American Corp.

ACME Signs & Displays - Post-Pro
Philadelphia, Pa.

Bachrach and Company - 1895 - 1917
San Francisco, Ca.

Baltimore Enamel and Novelty Co. - Pre-Pro
Baltimore, Md.
Largest producer of "Porcelain" trays.

H.D. Beach - 1901 - Present
J. Coshoncton, Oh.

Brilliant Mfg. Co. - 1910 - 1930
Philadelphia, Pa.
Listed on a L. Bergdoll Brewing Co. 12" round brass tray. Louis Bergdoll Brewing Co., Philadelphia, Pa., closed in 1920. Arced along the bottom face of this tray, in 1/2" embossed print, is stamped Philada. Pa.; a rather unusual way of abbreviating Philadelphia.

The Burdick Co. - 1929 -? Most trays 1930s - early 1940s
New York, NY.

Carpathian Silvertray Co. - Pre-Pro
Park Place, NY.
Listed on a Schlitz copper plated tray

Continental Can Co. - 1913 - Present
New York, NY. Headquarters (multi-branch locations)

Chromo Eng. Makers, Saml. Buckley & Co.
New York
Only listed on a Hubert Fischer Porcelain tray
    Hartford, Ct.

Eertels Metal Ware Co.
Kingston, Pa.
Listed on a round 13" Mt. Carbon tray from Pottsville, Pa.

Electro-Chemical Engraving Co., Inc.
Bronx, NY.
Did not make trays until after Prohibition, ended tray production around 1950.

R.G. Fenstermaker Adv. Specialties
Allentown, Pa.
Limited Post-Pro tray supply company. Listed on Bushkill tray, Easton, Pa.

General Brewers Supply Corp.
New York, NY.
Limited Post-Pro tray supply company listed on Lion Tray Wilkes-Barre, Pa.

Grammes, Inc. - 1875 - ?
Allentown, Pa.
Listed on Neuweilers Tray Allentown, Pa.

Haeusermann Litho - 1905 - 1921
NY and Chicago

The International AD Co.
NY
Listed on a Hubert Fischer Tray, enamel over tin, or porcelain composition (Hartford, Ct.) probably a supply company.

Kaufmann and Strauss Co. - 1890 - ?
New York, NY.
I believe this company survived Prohibition and made a small number of Post-Pro trays.

Ludlow Mfg. Co. - Post-Pro
Cincinnati, Oh.
This company is listed on a round 12" Student Prince Tray from Heidelberg Brewing Co. (Covington, Ky.) It is very shallow with a wide, flat upper rim.

F.E. Marsland - Pre-Pro
New York, NY.
Listed on 2 Hubert Fischer (Hartford, Ct.) trays, one porcelain, one etched, plated tin. Street address was W. Broadway. Probably a supply Company Baltimore Enamel and Novelty Co. also listed W. Broadway, NY, New York as a branch address.

Mason Can Co. - Post-Pro
Providence. RI.
Trays were never a major product for Mason. Some examples produced by Mason are trays made for Star (Murphy's Ale-Boston), Ehret's (Brooklyn), and Franklin (Wilkes-Barre, Pa.) I believe all trays were 13" round, straight sided.

Mayer and Lavenson Co. - 1903 - 1914
New York, NY.
Some examples include trays made for Feigenspan (Newark), Doelger (New York), and Consumers (New York).

The Meek Co. - 1901 - 1909
Coshoncton, Oh.
Manufactured a large number of early trays. Remained in business as American Art Works after Jasper F. Meek retired.

The Meek and Beach Co. - 1901 - 1905
Coshoncton, Oh.
This guide includes many trays from either Meek or Beach but few of which were produced during their short partnership. One is a Harvard Tray (Lowell, Ma), another is from the Hartmann Brewing Co. (Bridgeport, Ct.).

The Metallograph Corp. - Post-Pro
New York, NY.
More likely to be a supply company than a manufacturer. Very few trays bear the Metallograph Corp. name. Two examples are Tru-Age Standard (Scranton, Pa.), and New England Brewing Co. (Hartford, Ct.).

EAGLE BREWING COMPANY
OLD DUTCH PREMIUM BEER
CATASAUQUA, PA.
MFG: ?
ROUND 12"
1950s

The Nelke Co.
Listed as L.D. Nelke signs in the 1916 Thomas Register. Listed as the Nelke Sign Manufacturing Corp. in the 1962 McRaes. Very few trays are accredited to them. One example is a round 13" Pie for Cremo (New Britain, Ct.).

New York Importing Co. - Post-Pro
New York, NY.
Likely a supply company, subcontracting Burdick for tray manufacturing, majority of trays listing New York Importing Co. are round 13" pies.

N.Y. Metal Sign Works Co. - Pre-Pro
New York, NY.
Listed on a very shallow 13" pie tray from Falls City Brewing Co. Incorporated, Louisville, KY.

Niagra Litho Co. - Post-Pro
Buffalo, NY.
Trays include Iroquois (Buffalo, NY.) and Cataract (Rochester, NY.).

Ivan B. Nordham Company - Pre-Pro
Location unknown. Credited on a Pre-Pro tray for Pennsy Select Beer (Moose Brewing Co., Roscoe, Pa.)

Norton Bros., Press - Pre-Pro
Chicago, IL.
Listed on a 12" round Pabst tray, Milwaukee, WI, which is quite unusual in style. It is a thin gauge very deeply embossed tray, probably pre-1900.

The Novelty Advertising Co. - 1893 - Present
Coshoncton, Oh.
Almost all the round 14" trays shown are from The Novelty Advertising Co.

F.J. Offermann Art Works, Inc. - Limited Post-Pro
Buffalo, NY.
Two trays they manufactured are Kuhn's (Jamestown, NY.) and Phoenix (Buffalo, NY.), both round 12".

Owens-Illinois Can Company - Limited Post-Pro
Multiple locations
Listed in 1946 Thomas' Register as a division of Continental Can Co. Trays carrying their name include a Stegmaier (Wilkes-Barre, Pa) and a Linden (Lindenhurst, Long Island, NY.) A round 13" Harvard tray made of aluminum, with a polished, unpainted face lists its manufacturer as O.I.C. Co. which I assume is Owens-Illinois, since its size and shape characteristics are consistent with trays produced by them.

PACO - Pre-Pro
San Antonio, TX.
A small, limited output company credited on two 11" x 14" rectangular trays manufactured for the San Antonio Brewing Assn. (San Antonio, Texas). The brands advertised are Pearl and Texas Pride. I am not familiar with any PACO trays which are not rectangular 11" x 14" models.

P.F. and Co. (Palm, Fechtler, and Co.)
New York, NY.
Originally in New York but later relocated in Weehawken, NJ. Credited on Pre-Pro Ringler Tray (New York, NY.).

Passaic Metal Ware Co. Litho
Passaic, NJ.
Made a number of Pre-Pro beer trays as well as many early Coca-Cola trays. Listed in 1931 as a division of Continental Can Co.

Pilgrim Novelty Co.
Providence, RI.
A Pre-Pro Minster Tray (Consumers, Cranston, RI.) is the only tray included that bears the Pilgrim name.

Louis Roesch
San Francisco, Ca.
Very early, small Pre-Pro company. No trays included in this book are credited to Roesch. One tray credited to them is the very well known Oval National Tray (San Francisco, Ca.) with two cowboys on horseback entitled "Pastimes on the Frontier".

J.W. Russell & Co. - Pre-Pro
New York, NY.
This company produced a very unusual and quite attractive tray for the Providence Brewing Co., Providence, RI. It is a small oval brass tray, advertizing Bohemian, PB, Standard, Columbian, and Extra Lager brands. The tray's brass face is polished. Across its center, the Providence Brewing Co. is advertized in a black and red painted logo.

Sentenne and Green - 1894 - 1923
New York, NY.

Chas. W. Shonk Co. Litho - 1890 - late 1920s
Chicago, IL.
Shonk was one of the largest of the Pre-Pro tray manufacturing companies. They made round 12" straight sided trays, round 12" and 13" pies, 11" x 14" rectangular trays, and also ovals.

Chas. W. Shonk - American Can Co.
Haywood, Il.
It seems likely that Shonk was bought out by American Can Co. at some point during Prohibition. They are listed on the Congo near beer tin tray from Haberle-Crystal Bottling Dept. (Syracuse, NY.) made during Prohibition. This tray is the only example carrying the combined names of Shonk/American Can Co. included in this guide.

Standard Advertising Co. - 1891 - 1901
Coshoncton, Oh.
This company was H.D. Beach's original business before his merger with J.F. Meek's Tuscorora Advertising Co. The Stegmaier 13" pie Factory Scene (Wilkes-Barre, Pa.) is one Standard Advertising example.

Tindeco (The Tin Decorating Company of Baltimore)
One example is a rectangular Valley Forge tray from Adam Scheidt entitled "Washington's Headquarters". Whether due to material quality or manufacturing process, Tindeco Trays generally did not age particularly well. Tindeco Trays formerly used for their intended serving purposes often exhibit a faded dullness of finish, water marks, flaking, etc.

Tuscarora Advertising Co. - 1888 - 1901
Coshoncton, Oh.

Universal Tray and Sign Co. - Post-Pro
New York, NY.
Possibly a supply company. Credited on many 1930s trays. Most examples are round 12" straight sided types. (Poth's, Moore and Quinn, Erlanger's, Anthracite, Metzgers, etc). Some 13" pies (Krueger, (Newark)) were also produced.

Wolff and Lawrence - Pre-Pro
New York, NY.
Limited numbers of tray varieties were produced. One example is the Graham's Vitabrew Tray (Paterson, NJ) which is one of the few round 14" trays within that are not accredited to Novelty Advertising Co.

New Jersey

LEMBECK & BETZ, EAGLE BREWING CO.
LAGER BEERS, ALES & PORTER
JERSEY CITY, N.J.
CHAS. W. SHONK
RECT. 11" x 14"
PRE - PRO

P. BALLANTINE & SONS
BALLANTINE'S ALE - BEER
NEWARK, N.J.
A.C.CO.
ROUND 12"
1940s

THE Wm. PETER BREWING CO.
PETER BRAU SELECTED MALT AND HOPS BEER
UNION HILL, N.J.
HAEUSERMANN LITHO
OVAL 14" x 17"
c. 1905

P. BALLANTINE & SONS
BALLANTINE ALE & BEER
NEWARK, N.J.
MFG:
ROUND 12"
1950s

P. BALLANTINE & SONS
BALLANTINE BEER
NEWARK, N.J.
CCC
ROUND 13"
1960s

P. BALLANTINE & SONS
BALLANTINE ALES - BEERS
NEWARK, N.J.
AMERICAN CAN CO.
ROUND 12"
1950s

CAMDEN COUNTY BEVERAGE CO.
CAMDEN BEER
CAMDEN, N.J.
CCC
ROUND 12"
c. 1953 (song lyrics)

PETER BREIDT BREWING CO.
BREIDT'S BEER - ALE
ELIZABETH, N.J.
ELECTRO - CHEMICAL ENGRAVING CO., INC.
ROUND 13"
1938

CAMDEN COUNTY BEVERAGE CO.
CAMDEN BEER
CAMDEN, N.J.
CCC
ROUND 12"
1950s

CAMDEN COUNTY BEVERAGE CO
CAMDEN BEER
CAMDEN, N.J.
MFG: ?
ROUND 12"
EARLY 1940s

CAMDEN COUNTY BEVERAGE CO., INC
CAMDEN LAGER, PILSNER, LORD CAMDEN ALE
CAMDEN, N.J.
MFG: ?
ROUND 13" PIE
1930s

ELIZABETH BREWING CORP.
BREIDT'S BEERS - ALES
ELIZABETH, N.J.
ELECTRO - CHEMICAL ENGRAVING CO., INC
ROUND 12"
1935

CHRISTIAN FEIGENSPAN BREWERIES
FEIGENSPAN P.O.N
NEWARK, N.J.
THE MEEK CO.
ROUND 13" PIE
1901 - 1909

CHRISTIAN FEIGENSPAN, INC.
FEIGENSPAN P.O.N. PRIVATE SEAL BEER
NEWARK, N.J.
STELAD SIGNS - PASSAIC METAL WARE CO.
ROUND 13"
PRE - PRO

CHRISTIAN FEIGENSPAN BREWING CO.
FEIGENSPAN P.O.N
NEWARK, N.J.
A.C.CO. ARTIST A. ASTI
ROUND 13"
MID - LATE 1930s

CHRISTIAN FEIGENSPAN BREWERIES
NEWARK, N.J.
MAYER AND LAVINSON
ROUND 13" PIE
c. 1910

CHRISTIAN FEIGENSPAN BREWING CO.
FEIGENSPAN P.O.N.
NEWARK, N.J.
ELECTRO - CHEMICAL ENGRAVING CO., INC.
ROUND 13"
1930s

CHRISTIAN FEIGENSPAN BREWING CO.
FEIGENSPAN P.O.N
NEWARK, N.J.
MFG: ?
ROUND 12"
1930s

PETER HAUCK AND COMPANY
"BREWERS AND BOTTLERS OF EXCLUSIVE HIGH GRADE BEERS"
HARRISON, N.J.
KAUFMANN AND STRAUSS
"PURITY"
ROUND 12"
PRE - PRO

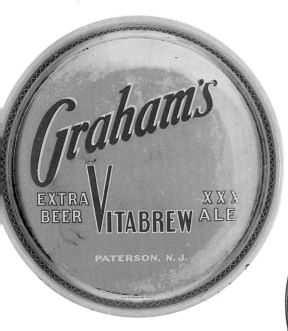

JAMES A. GRAHAM BREWING CO.
GRAHAM'S VITABREW, EXTRA BEER, XXXALE
PATERSON, N.J.
WOLFF AND LAWRENCE
ROUND 14"
PRE - PRO

HOFFMANN BEVERAGE CO.
HOFFMANN BEER
NEWARK, N.J.
MFG: ?
ROUND 13"
1934 - 1946 (possibly post '46 by Pabst and Hoffmann)

JAMES A. GRAHAM BREWING CO.
GRAHAM'S VITABREW, XXXALE
PATERSON, N.J.
MFG: ?
ROUND 12"
PRE - PRO

HOFFMANN BEVERAGE CO.
HOFFMANN BEER
NEWARK, N.J.
MFG: ?
ROUND 13"
1934 - 1946

JOSEPH HENSLER BREWING CO.
HENSLER'S BEER
NEWARK, N.J.
ELECTRO - CHEMICAL ENGRAVING CO., INC
ROUND 12"
1940s

JOSEPH HENSLER BREWING CO.
HENSLER PRIVATE LABEL PREMIUM DRY BEER
NEWARK, N.J.
MFG: ?
ROUND 12"
EARLY 1950s

THE JOSEPH HENSLER BREWING CO.
HENSLER LIGHT BEER, HENSLER PRIVATE LABEL
NEWARK, N.J.
AMERICAN CAN CO.
ROUND 12"
1950s

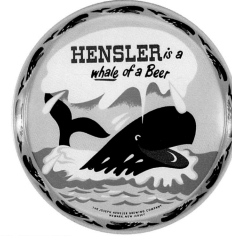

THE JOSEPH HENSLER BREWING COMPANY
HENSLER BEER
NEWARK, N.J.
MFG: ?
ROUND 12"
1950s

JOSEPH HENSLER BREWING CO.
HENSLER'S BEER & ALES
NEWARK, N.J.
ELECTRO - CHEMICAL ENGRAVING CO., INC.
ROUND 12"
1930s

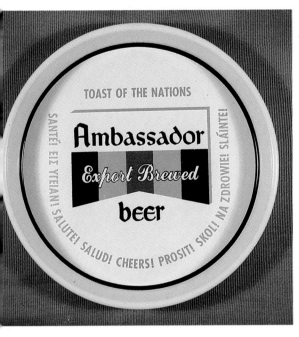

G. KRUEGER BREWING CO.
AMBASSADOR EXPORT BREWED BEER
NEWARK, N.J.
A.C.CO.
ROUND 12" AND 13" SIZES
MID-1950s

G. KRUEGER BREWING CO.
KRUEGER'S BEER AND ALES
NEWARK, N.J.
UNIVERSAL TRAY AND SIGN CO.
ROUND 13" PIE
MID - LATE 1930s

G. KRUEGER BREWING CO.
AMBASSADOR EXPORT BREWED BEER
NEWARK, N.J.
MFG: ?

G. KRUEGER BREWING CO.
KRUEGERS BEER - ALES
NEWARK, N.J.
AMERICAN CAN CO.
ROUND 12"
1940s

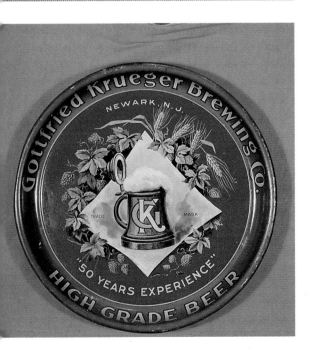

GOTTFRIED KRUEGER BREWING CO.
"HIGH GRADE BEER"
NEWARK, N.J.
AMERICAN ART WORKS
ROUND 13" PIE
1912 - 1918

G. KRUEGER BREWING CO.
KRUEGER BEER
NEWARK, N.J.
AMERICAN CAN CO.
ROUND 12"
1950s

G. KRUEGER BREWING CO.
KRUEGER BEER - ALE
NEWARK, N.J.
AMERICAN CAN CO.
ROUND 12"
1950s

KRUEGER BREWING CO. / FALSTAFF
KRUEGER PILSNER BEER
PROVIDENCE, (CRANSTON) R.I.
CCC
ROUND 12"
1965 - 1970

G. KRUEGER BREWING CO.
KRUEGER BEER - ALE
NEWARK, N.J.
A.C. CO.
ROUND 12"
LATE 1940s

G. KRUEGER BREWING CO.
KRUEGER ALE - BEER
NEWARK, N.J.
MFG: ?
ROUND 13"
EARLY 1950s

KRUEGER BREWING CO. / FALSTAFF
KRUEGER PILSNER BEER
PROVIDENCE,(CRANSTON) R.I.
CCC
ROUND 12"
1965 - 1970

* Note: The two Krueger Pilsner Beer trays above (ROUND 12", 1965-1970) were included with the other Krueger trays for grouping continuity purposes. They actually were made after Krueger no longer existed in Newark. The Krueger name was bought by Narragansett and later by Falstaff after Krueger closed in 1961, hence they actually advertise the Krueger brand name during the years it was brewed in Rhode Island.

PEOPLE'S BREWING CO.
TRENTON OLD STOCK BEER
TRENTON, N.J.
NOVELTY ADV. CO.
ROUND 14"
1930s

PEOPLE'S BREWING CO.
TRENTON OLD STOCK BEER, TRENT ALE
TRENTON, N.J.
UNIVERSAL TRAY AND SIGN
ROUND 12"
LATE '30s, EARLY '40s

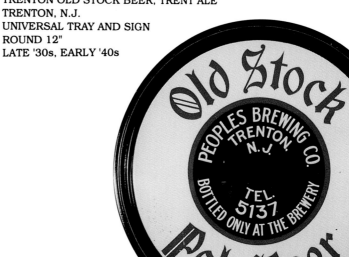

PEOPLE'S BREWING CO.PEOPLE'S BREWING COMPANY
TRENTON OLD STOCK BEEROLD STOCK PALE BEER
TRENTON, N.J.TRENTON, N.J.
A.C. CO.THE NOVELTY ADV. CO.
ROUND 12"ROUND 14"
LATE '30sMID - LATE '30s

THE Wm. PETER BREWING CORP.
PETER 1859 BEER
UNION CITY, N.J.
ELECTRO - CHEMICAL ENGRAVING CO., INC.
ROUND 12"
1940

THE WM. PETER BREWING CORP.
PETER 1859 BEER
UNION CITY, N.J.
ELECTRO-CHEMICAL ENGRAVING CO., INC.
ROUND 12"
1940

THE Wm. PETER BREWING CORP.
PETER BEER, ALES, PORTER
UNION CITY, N.J.
ELECTRO - CHEMICAL ENGRAVING CO., INC.
ROUND 12"
1937

RISING SUN BREWING CO.
SEEBER SPECIAL
ELIZABETH, N.J.
KAUFMANN AND STRAUSS
ROUND 12"
PRE - PRO

RISING SUN BREWING CO.
SEEBER BREW - BOHEMIA BREW
ELIZABETH, N.J.
KAUFMANN AND STRAUSS
ROUND 12"
PRE - PRO

SCHULTZ BREWING CO., INC.
SCHULTZ BEER - ALE
UNION CITY, N.J.
BURDICK
ROUND 12"
1934 - 1938

UNION BREWING CO of NEWARK
NEWARK, N.J.
H. D. BEACH CO.
ROUND 12"
1901 - 1919

SCHULTZ BREWING CO., INC.
OLD LONDON ALE & BEER
UNION CITY, N.J.
BURDICK
ROUND 12"
1930s

UNITED BREWING COMPANY
OLDBURGER BEER
NEWARK, N.J.
MFG: ?
ROUND 12"
1933 - 1938

SEEBER BREWING CO.
SEEBER BREWS
ELIZABETH, N.J.
AMERICAN ART WORKS
ROUND 13" PIE
1933 - 1937

Massachusetts

BROCKERT BREWING CO., INC.
BROCKERT'S 3XXX ALE
WORCESTER, MASS.
UNIVERSAL TRAY AND SIGN
ROUND 13" PIE
1935 - 1945

TOBIN'S CREAMY ALE
I HAVE BEEN UNABLE TO FIND MUCH INFORMATION ABOUT
THIS TRAY. DOES ANYONE KNOW ANY DETAILS ABOUT TOBIN?
UNIVERSAL TRAY AND SIGN
ROUND 12"

A.G. VAN NOSTRAND BUNKER HILL BREWERIES
P.B. ALE, LAGER, PORTER
BOSTON, MASS.
CHAS. W. SHONK
ROUND 12"
c. 1906

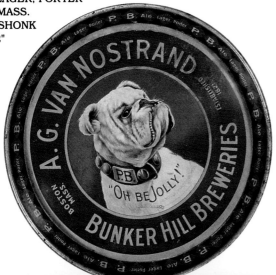

BROCKERT BREWING CO., INC.
BROCKERT'S ALE
WORCESTER, MASS.
MFG: ?
ROUND 13" PIE
LATE 1930s, EARLY 1940s

BURKHARDT BREWING COMPANY
TIVOLI, STOCK ALE, CREAM ALE, AUGSBURGER
BOSTON, MASS.
KAUFMANN AND STRAUSS
ROUND 12"
1910 - 1918

COLD SPRING BREWING CO.
LAWRENCE, MASS.
THE H. D. BEACH CO.
OVAL 14" x 17"
1900 - 1910

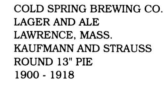

COLD SPRING BREWING CO.
COLD SPRING ALE AND BEER
LAWRENCE, MASS.
THE BEACH CO.
ROUND 13"
1930s

COMMONWEALTH BREWING CORPORATION
OXFORD CLUB ALE
SPRINGFIELD, MASS.
THE BURDICK CO.
ROUND 13" PIE
MID - LATE 1930s

COLD SPRING BREWING CO.
LAGER AND ALE
LAWRENCE, MASS.
KAUFMANN AND STRAUSS
ROUND 13" PIE
1900 - 1918

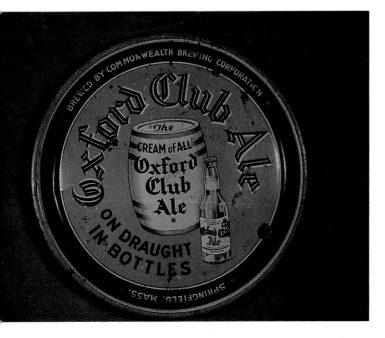

COMMERCIAL BREWING CO.
OLD INDIA PALE ALE
BOSTON, MASS.
ROUND 12"
1933 - 1940

COMMERCIAL BREWING CO.
OLD INDIA PALE ALE
MFG: ?
ROUND 12"
1933 - 1940

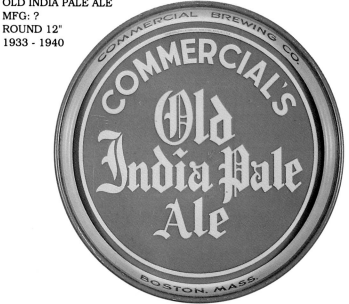

CROFT BREWING CO.
CROFT ALE
BOSTON, MASS.
AMERICAN CAN CO.
ROUND 12"
LATE 1930s, 1940s

CROFT BREWING CO. (possibly Narragansett)
CROFT CREAM ALE
BOSTON, MASS.
CANCO
ROUND 12"
1950 - 1953

DAWSON'S BREWERY, INC.
DAWSON'S ALE & LAGER
NEW BEDFORD, MASS.
H. D. BEACH CO.
ROUND 12"
MID-1930s

CROFT BREWING CO.
CROFT ALE
BOSTON, MASS.
H. D. BEACH
ROUND 12"
LATE 1930s

CROFT BREWING CO. DAWSON'S BREWERY, INC.
PILGRIM ALE DAWSON'S GOLD CROWN ALE
BOSTON, MASS. NEW BEDFORD, MASS.
H. D. BEACH MASON CAN CO.
ROUND 12" ROUND 13"
LATE 1930s 1940s

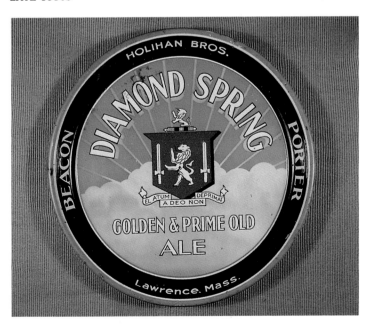

DAWSON'S BREWERY, INC.
DAWSON'S ALE & BEER
NEW BEDFORD, MASS.
A.C.CO.
ROUND 12"
EARLY 1940s

DIAMOND SPRING BREWERY (Holihan Bros)
GOLDEN & PRIME OLD ALE, BEACON, PORTER
LAWRENCE, MASS.
MFG: ?
ROUND 13" PIE
1940s

DIAMOND SPRING BREWERY
HOLIHAN BROTHERS
DIAMOND SPRING ALE, HOLIHAN'S LAGER
LAWRENCE, MASS.
MFG: ?
ROUND 12"
1930s

DIAMOND SPRING BREWERY, HOLIHAN BROTHERS
DIAMOND SPRING ALE, LAGER, PORTER
LAWRENCE, MASS.
MFG: ?
ROUND 13" PIE
1912 - 1918

DIAMOND SPRING BREWERY, HOLIHAN BROS.
DIAMOND SPRING ALE, LAGER, STOCKALE, PORTER
LAWRENCE, MASS.
KAUFMANN & STRAUSS
ROUND 13" PIE
1912 - 1918

DIAMOND SPRING BREWERY
HOLIHAN'S ALE
LAWRENCE, MASS.
AMERICAN ART WORKS
ROUND 12"
1940s

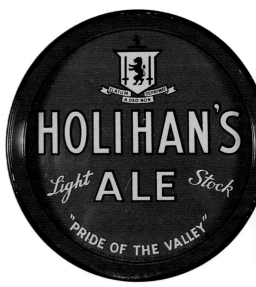

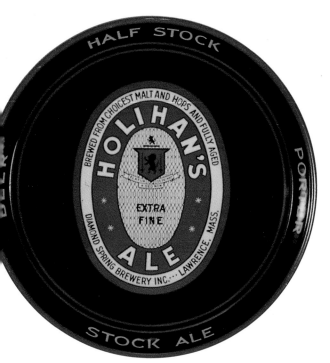

DIAMOND SPRING BREWERY, INC.
HOLIHAN'S ALE, HALF STOCK,
STOCK ALE, BEER, PORTER
LAWRENCE, MASS.
MFG: ?
ROUND 12"
1930s

ENTERPRISE BREWING CO.
OLD TAP ALES, BOHEMIAN BEER
FALL RIVER, MASS.
MFG: ?
ROUND 12"
1950s

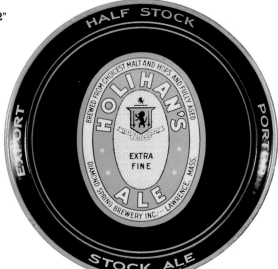

DIAMOND SPRING BREWERY, INC.
HOLIHAN'S ALE, HALF STOCK,
STOCK ALE, BEER, PORTER
LAWRENCE, MASS.
MFG: ?
ROUND 12"
1940s

ENTERPRISE BREWING CO.
BOHEMIAN LAGER
FALL RIVER, MASS.
MFG: ?
RECT. 11" x 14"
possibly PRE - PRO, if not,
MID-1930s

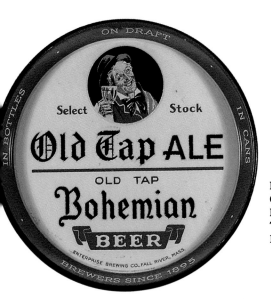

ENTERPRISE BREWING CO.
OLD TAP ALE, BOHEMIAN BEER
FALL RIVER, MASS.
THE BEACH CO.
ROUND 13"
1940s

ENTERPRISE BREWING CO.
BOHEMIAN BEER, OLD TAP ALES
FALL RIVER, MASS.
MFG: ?
ROUND 12"
1950s

ENTERPRISE BREWING CO.
OLD TAP ALE, BOHEMIAN BEER
FALL RIVER, MASS.
AMERICAN ART WORKS
ROUND 12"
1941

HAFFENREFFER & CO.
BOYLESTON ALE & LAGER
BOSTON, MASS.
HAEUSERMANN LITHO
OVAL 14" x 17"
1905 - 1918

ENTERPRISE BREWING CO.
GOLDEN SPARKLING ALE, BOHEMIAN LAGER
FALL RIVER, MASS.
BALTIMORE ENAMEL AND NOVELTY CO.
ROUND 12" (PORCELAIN)
1900 - 1910

HAFFENREFFER & CO., INC.
PICKWICK ALE
BOSTON, MASS.
MFG: ?
ROUND 12"
1940s

HAFFENREFFER & CO., INC.
PICKWICK ALE
BOSTON, MASS.
MFG: ?
ROUND 12"
1940s

HAFFENREFFER & CO., INC.
PICKWICK ALE
BOSTON, MASS.
MFG: ?
ROUND 12"
1940s

HAMPDEN BREWING CO.
HAMPDEN ALE - BEER
WILLIMANSETT, MASS.
MFG: ?
ROUND 13" PIE
1940s

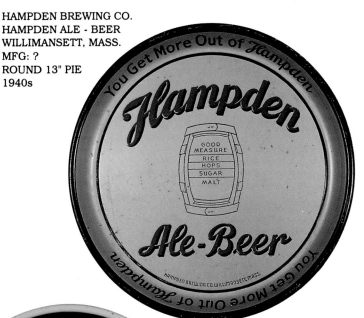

HAMPDEN BREWING CO.
HAMPDEN MILD ALE
WILLIMANSETT, MASS.
MFG: ?
ROUND 13" PIE
MID - LATE 1940s

HAMPDEN BREWING CO.
HAMPDEN MILD ALE
WILLIMANSETT, MASS.
CANCO
ROUND 13"
LATE 1940s, 1950s

HARVARD BREWING CO.
HARVARD ALE, EXPORT BEER,
CLIPPER ALE, PORTER
LOWELL, MASS.
O.I.C. CO.
ROUND 13" (polished aluminum)
1940

HARVARD BREWING CO.
HARVARD ALE, BEER, PORTER
LOWELL, MASS.
THE BEACH CO.
ROUND 13" (aluminum)
1940s

HARVARD BREWING CO.
HARVARD ALE, EXPORT BEER, PORTER
LOWELL, MASS.
THE BEACH CO
ROUND 13" (aluminum)
1940s

HARVARD BREWING CO.
HARVARD ALES - BEER - PORTER
LOWELL, MASS.
THE H. D. BEACH CO.
ROUND 12"
LATE 1930s, EARLY 1940s

HARVARD BREWING CO.
HARVARD ALE, BEER
LOWELL, MASS.
MFG: ?
ROUND 12"
1940s

HARVARD BREWING CO.
CLIPPER ALE
LOWELL, MASS.
A.C. CO.
ROUND 13"
1940s

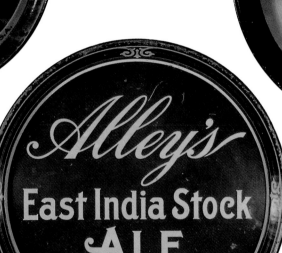

HARVARD BREW. CO.
HARVARD BEER
LOWELL, MASS.
THE MEEK AND BEACH CO.
ROUND 13" PIE
1901 - 1905

MASSACHUSETTS BREWERIES COMPANY
ALLEY'S EAST INDIA STOCK ALE
BOSTON, MASS.
KAUFMANN AND STRAUSS
ROUND 12"
1901 - 1918

MASSACHUSETTS BREWERIES COMPANY
ABC LAGER BEER
BOSTON, MASS.
CHAS. W. SHONK
ROUND 12" PIE
1901 - 1918

HARVARD BR'G CO.
HARVARD EXPORT GREEN LABEL BEER
LOWELL, MASS.
H. D. BEACH CO.
ROUND 12"
probably 1930s

45

MASSACHUSETTS BREWERIES COMPANY
KING'S BOHEMIAN
BOSTON, MASS.
KAUFMANN AND STRAUSS
ROUND 13" PIE
1901 - 1919

SMITH BROS.
SMITH BROS. ALES
NEW BEDFORD, MASS.
THE BALTIMORE ENAMEL AND NOVELTY CO.
ROUND 12" PIE (Porcelain)
1900 - 1910

MASSACHUSETTS BREWERIES COMPANY
PFAFF'S LAGER
BOSTON, MASS.
CHAS. W. SHONK
ROUND 12" PIE
1901 - 1920

SMITH BROS., INC.
SMITH BROS. ALES AND LAGER
NEW BEDFORD, MASS.
A.C. CO.
ROUND 12"
LATE 1930s, 1940s

MASSACHUSETTS BREWERIES COMPANY
PFAFF'S BEER
BOSTON, MASS.
MFG: ?
ROUND 12"
1901 - 1918

SPRINGFIELD BREWERIES CO.
HAMPDEN, HIGHLAND, TIVOLI, WURZBURGER
SPRINGFIELD, MASS.
CHAS. W. SHONK
ROUND 12" PIE
PRE - PRO

STAR BREWING CO.
STAR ALES AND LAGER
BOSTON, MASS.
UNIVERSAL TRAY AND SIGN
ROUND 12"
MID - LATE 1930s

SPRINGFIELD BREWERIES CO.
SPRINGFIELD BEER
SPRINGFIELD, MASS.
MFG: ?
ROUND 13" PIE
1910 - 1918

STAR BREWING CO.
STAR ALES AND LAGER
BOSTON, MASS.
A.C. CO.
ROUND 12"
1940s

STAR BREWING CO.
STAR PRIZE ALES AND LAGER
BOSTON, MASS.
MFG: ?
ROUND 13" PIE
PRE - PRO

STAR BREWING CO.
STAR PRIZE ALES AND LAGER
BOSTON, MASS.
THE MEEK CO. STOCK #72
"HEARTS ARE TRUMP"
RECT. 13" x 18"
1901 - 1909

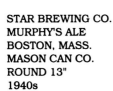

STAR BREWING CO.
MURPHY'S ALE
BOSTON, MASS.
MASON CAN CO.
ROUND 13"
1940s

WORCESTER BREWING CORPORATION
ALES - LAGER - PORTER
WORCESTER, MASS.
CHAS. W. SHONK
ROUND 12"
1900 - 1910

THE WORCESTER BREWING COMPANY
TADCASTER BEER AND ALE
WORCESTER, MASS.
AMERICAN ART WORKS
ROUND 12"
LATE 1940s

THE WORCESTER BREWING COMPANY
TADCASTER BEER AND ALE
WORCESTER, MASS.
AMERICAN ART WORKS
ROUND 12"
LATE 1940s

Rhode Island

WHAT CHEER BREWERY (MOLTERS)
PRINCESS ALE, WHAT CHEER LAGER
PROVIDENCE, R.I.
KAUFMANN AND STRAUSS
ROUND 13" PIE
1910

THE EAGLE BREWING CO.
RED STAR ALE
PROVIDENCE, R.I.
MFG: ?
ROUND 12"
MID - LATE 1890s

CONSUMERS' BREWING CO.
MINSTER LAGER BEER, CONSUMERS ALE
CRANSTON, R.I.
PILGRIM NOVELTY CO.
ROUND 12"
1911 - 1920

CONSUMER'S BREWERY CO., INC.
CONSUMER'S BEER
HILLSGROVE, WARWICK, R.I.
MFG: ?
ROUND 13" PIE
1933 - 1935

THE JAMES HANLEY BREWING CO.
HANLEY'S PEERLESS ALE
PROVIDENCE, R.I.
BALTIMORE ENAMEL AND NOVELTY CO.
ROUND 12" PIE (porcelain)
1900 - 1910

EAGLE BREWING CO.
ALE AND LAGER
PROVIDENCE, R.I.
BALTIMORE ENAMEL AND NOVELTY CO.
OVAL 14" x 17"
EARLY 1900s

HAND BREWING CO.
HALF STOCK & LAGER
PAWTUCKET, R.I.
AMERICAN ART WORKS
STOCK #77
ROUND 13" PIE
1909 - 1919

THE JAMES HANLEY COMPANY
HANLEY'S PEERLESS ALE
PROVIDENCE, R.I.
MFG: ?
"THE CONNOISSEUR"
ROUND 13" PIE
EARLY 1930s

THE JAMES HANLEY COMPANY
HANLEY'S PEERLESS ALE
PROVIDENCE, R.I.
AMERICAN ART WORKS
ROUND 12"
EARLY 1940s

THE JAMES HANLEY COMPANY
HANLEY'S PEERLESS ALE
PROVIDENCE, R.I.
AMERICAN ART WORKS
ROUND 12"
EARLY 1940s

NARRAGANSETT BREWING CO.
SELECT STOCK LAGER, BANQUET ALE
PROVIDENCE, R.I.
BALTIMORE ENAMEL AND NOVELTY CO.
ROUND 12" PIE (porcelain)
1905 - 1910

THE JAMES HANLEY COMPANY
HANLEY'S PEERLESS ALE
PROVIDENCE, R.I.
STANDARD ADVERTISING CO.
ROUND 13" PIE
1896 - 1901

NARRAGANSETT BREWING COMPANY
NARRAGANSETT ALE, LAGER, MALT EXTRACT
PROVIDENCE, R.I.
THE MEEK COMPANY
OVAL 14" x 17"
1901 - 1909

THE JAMES HANLEY COMPANY
HANLEY'S BEER AND ALE
PROVIDENCE, R.I.
MFG: ?
ROUND 12"
LATE 1950s, EARLY 1960s

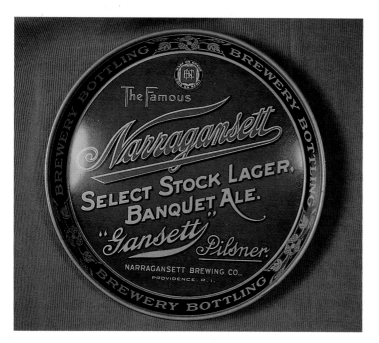

NARRAGANSETT BREWING CO.
SELECT STOCK LAGER, BANQUET ALE, "GANSETT" PILSNER
PROVIDENCE, R.I.
CHAS. W. SHONK
ROUND 12"
PRE - PRO

NARRAGANSETT BREWING CO.
SELECT STOCK LAGER AND BANQUET ALE
PROVIDENCE, R.I.
THE MEEK CO.
OVAL 14" x 17"
1907

NARRAGANSETT BREWING CO.
NARRAGANSETT LAGER AND ALE
PROVIDENCE, R.I.
CHAS. W. SHONK
ROUND 12"
1910 - 1915

NARRAGANSETT BREWING CO
NARRAGANSETT LAGER AND ALE
PROVIDENCE, R.I.
MFG: ?
ARTIST: DR. SUESS
ROUND 12"
1940s

NARRAGANSETT BREWING CO.
NARRAGANSETT LAGER, ALE
PROVIDENCE, R.I.
THE BURDICK CO.
ROUND 12"
1930s

NARRAGANSETT BREWING CO.
NARRAGANSETT LAGER, ALE
PROVIDENCE, R.I.
THE BURDICK CO.
ROUND 12"
1930s

PROVIDENCE BREWING CO.
BOHEMIAN BEER AND CANADA MALT ALE
PROVIDENCE, R.I.
KAUFMANN AND STRAUSS
ROUND 13" PIE
PRE - PRO

NARRAGANSETT BREWING CO.
BANQUET ALE
PROVIDENCE, R.I.
MFG: ?
ROUND 13" PIE
1930s

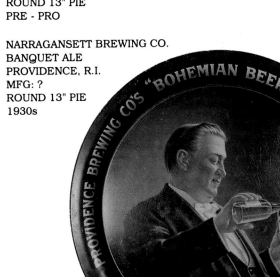

PROVIDENCE BREWING CO.
BOHEMIAN BEER AND CANADA MALT ALE
PROVIDENCE, R.I.
THE MEEK CO.
"A GOOD JUDGE" STOCK #97
ROUND 13" PIE
1903

NARRAGANSETT BREWING CO.
NARRAGANSETT ALE
PROVIDENCE, R.I.
THE BURDICK CO.
ROUND 12"
LATE 1930s, 1940s

PROVIDENCE BREWING CO.
PROVIDENCE ALE AND LAGER
PROVIDENCE, R.I.
THE BALTIMORE ENAMEL AND NOVELTY CO.
OVAL 13" x 16"
EARLY 1900s

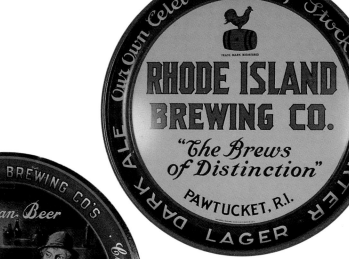

PROVIDENCE BREWING CO.
BOHEMIAN BEER
PROVIDENCE, R.I.
THE MEEK CO.
"READING THE NEWS" STOCK #19
OVAL 14" x 17"
1901 - 1909

PROVIDENCE BREWING CO.
BOHEMIAN BEER, CANADA MALT ALE
PROVIDENCE, R.I.
THE MEEK CO.
"FALSTAFF" STOCK #42
RECT. 12" x 17"
1901 - 1909

PROVIDENCE BREWING CO.
BOHEMIAN BEER - CANADA MALT ALE
PROVIDENCE, R.I.
THE MEEK CO.
"OLD FRIEND" STOCK #45
ROUND 13" PIE
1901 - 1909

RHODE ISLAND BREWING CO.
HALF STOCK ALE, DARK ALE, PORTER, LAGER
PAWTUCKET, R.I.
UNIVERSAL TRAY AND SIGN CO.
ROUND 13" PIE
1933 - 1939

New York

S. BOLTON'S SONS
HOME BREWED ALE
TROY, N.Y.
CHAS. W. SHONK
OVAL 14" x 19"
PRE - PRO

ONEIDA BREWING CO.
ONEIDA ALES AND PORTER
UTICA, N.Y.
CHAS. W. SHONK
"SHENANDOAH CHIEF OF ONEIDAS"
ROUND 12"
PRE - PRO

JOHN F. TROMMER'S EVERGREEN BREWERY
TROMMER'S EVERGREEN BEER
BROOKLYN, N.Y.
THE MEEK CO.
RECT. 13" x 18"
1902 - 1909

AMERICAN BREW. CO.
LIBERTY ROCHESTER BEER
ROCHESTER, N.Y.
HAEUSERMANN LITHO
ROUND 12"
1905 - 1920

AMERICAN BREWING CO.
TAM O'SHANTER LAGER BEER AND ALES
ROCHESTER, N.Y.
A.C.CO.
ROUND 13"
1933

AMERICAN BREWING CO.
APOLLO BEER, TAM O'SHANTER ALE
ROCHESTER, N.Y.
A.C.CO.
ROUND 12"
1930s, EARLY 1940s

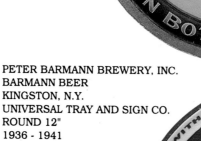

PETER BARMANN BREWERY, INC.
BARMANN BEER
KINGSTON, N.Y.
UNIVERSAL TRAY AND SIGN CO.
ROUND 12"
1936 - 1941

PETER BARMANN
THURINGER HOFBRAU, HALF STOCK ALE
KINGSTON, N.Y.
KAUFMANN & STRAUSS
ROUND 12"
1901 - 1908

BARTELS BREWING CO.
BARTELS
SYRACUSE, N.Y.
CHAS. W. SHONK
"NACHWACHTER" (NIGHT WATCHMAN)
ROUND 12"
1907 - 1910

BARTHOLOMAY BREWING CO.
BARTHOLOMAY BEERS, ALES, PORTER
ROCHESTER, N.Y.
CHAS. W. SHONK
ROUND 12"
1905

BEVERWYCK BREWING CO.
BEVERWYCK LAGER (LIGHT OR DARK)
ALBANY, N.Y.
AMERICAN ART WORKS
"THE INVITATION" W.H. McENTEE (artist)
RECT. 11" X 14"
C 1911

BEVERWYCK BREWERIES, INC.
ALBANY, N.Y.
NEW YORK IMPORTING CO.
ROUND 13" PIE
MID - LATE 1930s

BEVERWYCK BREWERIES, INC.
FAMOUS BEVERWYCK BEER
ALBANY, N.Y.
ELECTRO - CHEMICAL ENGRAVING CO., INC.
ROUND 12"
LATE 1930s

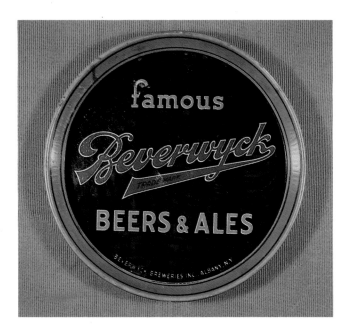

BEVERWYCK BREWERIES, INC.
FAMOUS BEVERWYCK BEERS - ALES
ALBANY, N.Y.
ELECTRO - CHEMICAL ENGRAVING CO., INC.
ROUND 12"
1935

BEVERWYCK BREWERIES, INC.
BEVERWYCK BEER
ALBANY, N.Y.
MFG: ?
ROUND 13"
1940s

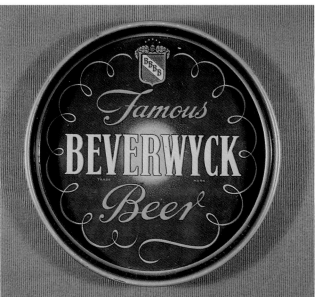

BEVERWYCK BREWERIES, INC.
FAMOUS BEVERWYCK BEER
ALBANY, N.Y.
ELECTRO - CHEMICAL ENGRAVING CO., INC.
ROUND 12"
1941

BROADWAY BREWING CO.
PURE BEERS
BUFFALO, N.Y.
HAEUSERMANN LITHO CO.
ROUND 12"
1905 - 1920

BEVERWYCK BREWERIES, INC.
BEVERWYCK BEERS & ALES
ALBANY, N.Y.
THE BURDICK CO., INC.
ROUND 13"
1930s

CATARACT BREWING CO., INC.
CANANDAIGUA BEER AND ALE
ROCHESTER, N.Y.
MFG: ?
OVAL 14" x 17"
1930s

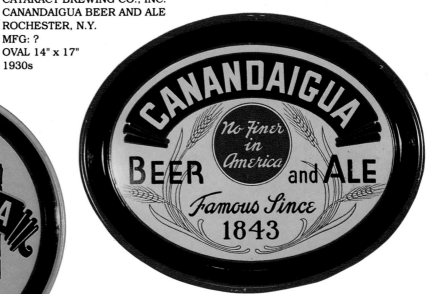

CATARACT BREWING CO., INC.
CANANDAIGUA HIGH HOPPED ALE
ROCHESTER, N.Y.
AMERICAN CAN CO.
ROUND 12"
1933 - 1940

E. & J. BURKE, LTD.
BURKE'S STOUT AND ALE
LONG ISLAND CITY, N.Y.
A.C. CO.
ROUND 12"
1934

BURKE BREWERY, INC.
GUINESS FOREIGN STOUT (under contract)
LONG ISLAND CITY, N.Y.
H. D. BEACH CO.
ROUND 12"
1940s

BUFFALO CO - OPERATIVE BREWING CO.
EXTRA 6 EXPORT LAGER, SUPERIOR XXX STOCK ALE
BUFFALO, N.Y.
MFG: ?
ROUND 13"
PRE - PRO

CATARACT BREWING CO., INC.
CATARACT CREAM ALE - LAGER BEER
ROCHESTER, N.Y.
NIAGARA LITHO CO. (Buffalo 2515)
ROUND 12"
1933 - 1940

CENTRAL BREWING CO.
68th ST. & EAST RIVER
NEW YORK, N.Y.
MAYER AND LAVENSON
"UNITY & PROGRESS" (Motto on wall)
OVAL 14" X 17"
1903 - 1914

CATARACT BREWING CO., INC.
CATARACT LAGER BEER
ROCHESTER, N.Y.
NIAGARA LITHO CO.
ROUND 12"
1933 - 1940

CITY BREWING CORPORATION
KOENIGS SPECIAL
NEW YORK, N.Y.
A.C. CO.
ROUND 13"
1933 - 1941

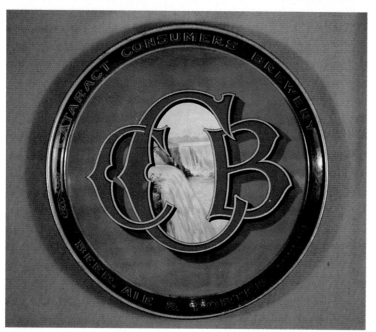

CATARACT CONSUMERS BREWERY
BEER, ALE & PORTER
NIAGARA FALLS, N.Y.
CHAS. W. SHONK LITHO
ROUND 12"
1905 - 1913

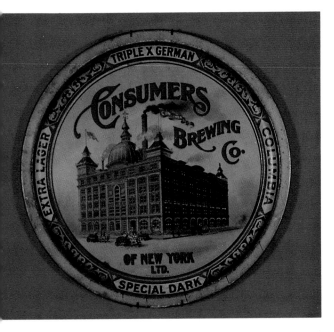

CONSUMERS BREWING CO. OF NEW YORK LTD
TRIPLEX GERMAN, EXTRA LAGER, COLUMBIA, SPECIAL DARK
NEW YORK, N.Y.
MAYER AND LAVENSON
ROUND 13" PIE
1911

THE DEER PARK BREW. CO.
LAGER BEER, ALE & PORTER
PORT JERVIS, N.Y.
H. D. BEACH (STOCK No. 7)
OVAL 14" x 17"
PRE - PRO

CRYSTAL SPRING BREW CO.
SUPERIOR STOCK LAGER, PALE BOHEMIAN, WURZBURGER
SYRACUSE, N.Y.
CHAS. W. SHONK
ROUND 12" PIE
1900 +/-

DEER PARK BEVERAGES, INC.
DEER PARK BEER
PORT JERVIS, N.Y.
UNIVERSAL TRAY AND SIGN CO.
ROUND 12"
1936 - 1942

DEER PARK BREWERIES, INC.
BEER - ALE - PORTER
PORT JERVIS, N.Y.
GENERAL BREWERS SUPPLY CORP.
ROUND 13" PIE
1933 - 1935

DOBLER BREWING CO.
DOBLER BEER AND ALE
ALBANY, N.Y.
AMERICAN CAN CO.
ROUND 12"
MID-1950s

PETER DOBLER, NEW YORK BREWERY
PETER DOBLER LAGER BEER
NEW YORK, N.Y.
MAYER AND LAVENSON
OVAL 14" x 17"
1903 - 1912

DOBLER BREWING CO.
DOBLER LAGER "A FAMOUS BREW"
ALBANY, N.Y.
CHAS. W. SHONK
OVAL 16" x 19"
1908 - 1919

PETER DOBLER BREWING CO.
PETER DOBLER BEER
NEW YORK, N.Y.
ELECTRO - CHEMICAL ENGRAVING CO., INC.
ROUND 12"
1933 - 1937

PETER DOBLER, NEW YORK BREWERY
PETER DOBLER FIRST PRIZE BEER
NEW YORK, N.Y.
HAEUSERMANN LITHO
OVAL 14" x 17"
1905 - 1921

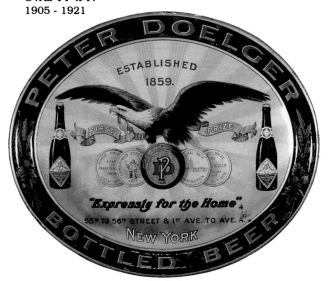

DOTTERWEICH BREWING CO.
GILT EDGE EXPORT LAGER - ALE - PORTER
OLEAN, N.Y.
THE MEEK CO.
(note: Apparently the artist felt that an ashtray for the cigar would clutter up the table too much)
ROUND 13" PIE
1901 - 1909

EAST BUFFALO BREWING CO.
LAGER BEER
BUFFALO, N.Y.
BALTIMORE ENAMEL AND NOVELTY CO.
OVAL 14" x 17" (porcelain)
EARLY 1900s

THE EAGLE BREWING CO.
EAGLE BEER
UTICA, N.Y.
NEW YORK IMPORTING CO.
ROUND 13" PIE
MID-1930s

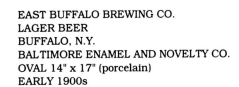

THE EAGLE BREWING CO.
PEERLESS LAGER
UTICA, N.Y.
HAEUSERMANN LITHO CO.
ROUND 12"
1913 - 1919

THE EAGLE BREWING CO.
EAGLE ALES AND LAGER
UTICA, N.Y.
AMERICAN CAN CO.
ROUND 12"
LATE 1930s, EARLY 1940s

THE EBLING BREWING CO., INC.
EBLING'S WHITE HORSE ALE
NEW YORK, N.Y.
ELECTRO - CHEMICAL ENGRAVING CO., INC.
ROUND 13"
MID-1930s

EDELBRAU BREWERY INC
EDELBRAU BEER
BROOKLYN, N.Y.
THE BURDICK CO., INC.
ROUND 13"
MID - LATE 1930s

THE EBLING BREWING CO., INC.
EBLING'S EXTRA
NEW YORK, N.Y.
NEW YORK IMPORTING CO.
ROUND 13"
LATE 1930s

THE EBLING BREWING CO., INC.
EBLING'S BEER & ALE
NEW YORK, N.Y.
ELECTRO - CHEMICAL ENGRAVING CO., INC.
ROUND 13"
1934

THE EBLING BREWING CO., INC.
EBLING'S BEER AND ALE
NEW YORK, N.Y.
ELECTRO - CHEMICAL ENGRAVING CO., INC.
ROUND 13"
1942

GEORGE EHRET BREWERY, INC.
EHRET'S EXTRA BEER
BROOKLYN, N.Y.
MASON CAN CO.
ROUND 13"
1940s

GEORGE EHRET HELL GATE BREWERY
GEO. EHRET'S EXTRA
NEW YORK, N.Y.
THE H. D. BEACH CO.
ROUND 12"
PRE - PRO

THE JOHN EICHLER BREWING CO.
EICHLER BEER
NEW YORK, N.Y.
ELECTRO - CHEMICAL ENGRAVING CO., INC.
ROUND 12"
1940

THE JOHN EICHLER BREWING CO.
EICHLER BEERS
NEW YORK, N.Y.
HAEUSERMANN LITHO CO.
ROUND 12"
1915

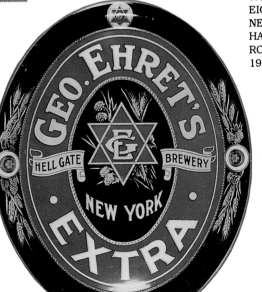

GEO. EHRET'S HELL GATE BREWERY
EHRET'S EXTRA
NEW YORK, N.Y.
H. D. BEACH CO.
OVAL 14" x 17"
PRE - PRO

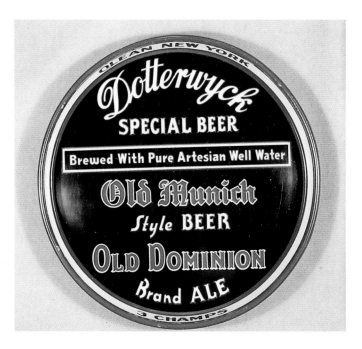

EMPIRE STATE BREWERY CORP.
DOTTERWYCK SPECIAL, OLD MUNICH, OLD DOMINION ALE
OLEAN, N.Y.
NOVELTY ADV. CO.
ROUND 12"
1933 - 1940

FIDELIO BREWERY INC.
McSORLEY'S LAGER BEER, CREAM STOCK ALE
NEW YORK, N.Y.
MFG: ?
ARTIST: WALTER BEACH HUMPHREY
RECT. 11" x 14"
CO. 1935

EMPIRE STATE BREWERY CORP.
DOTTERWYCK BEER - ALE
OLEAN, N.Y.
AMERICAN ART WORKS
ROUND 12"
1933 - 1940

FIDELIO BREWERY
FIDELIO BEER
NEW YORK, N.Y.
AMERICAN CAN CO.
ROUND 12"
1933 - 1940

FIDELIO BREWERY
FIDELIO BEER AND ALE
NEW YORK, N.Y.
MFG: ?
ROUND 12"
CO. 1936 (Painted by Herbert Bohnert)

FITZGERALD BROS. BREWING COMPANY
FITZGERALD'S ALE - BEER
TROY, N.Y.
A.C.CO.
ROUND 13"
EARLY - MID 1930s

SYRACUSE BREWERY, INC. (affiliated with Genesee)
DICKENS ALE
SYRACUSE, N.Y.
AMERICAN CAN CO.
ROUND 12"
1934 - 1937

FLANAGAN - NAY BREWING CO.
FLANAGAN - NAY "IT'S A MAN'S BEER"
BROOKLYN, N.Y.
ELECTRO - CHEMICAL ENGRAVING CO., INC.
ROUND 13"
1933 - 1937

GENESEE BREWING CO., INC.
GENESEE 12 HORSE ALE
ROCHESTER, N.Y.
A.C. CO.
ROUND 13"
1930s

GENESEE BREWING CO., INC.
GENESEE 12 HORSE ALE
ROCHESTER, N.Y.
MFG: ?
ROUND 13" PIE
1930s

GENESEE BREWING CO., INC.
LIEBOTSCHANER ALL MALT BEER
ROCHESTER, N.Y.
A.C. CO.
ROUND 12"
1930s

GENESEE BREWING CO., INC.
GENESEE BEER
ROCHESTER, N.Y.
CANCO
ROUND 12"
1950s

GENESEE BREWING CO., INC.
GENESEE ALE - BEER
ROCHESTER, N.Y.
MFG: ?
ROUND 12"
EARLY 1940s

GENESEE BREWING CO., INC.
GENESEE BEER
ROCHESTER, N.Y.
CANCO
ROUND 12"
1950s

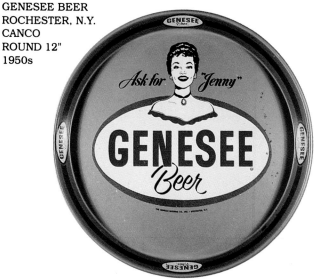

GENESEE BREWING CO., INC.
GENESEE ALE - BEER
ROCHESTER, N.Y.
CANCO
ROUND 12"
LATE 1940s

GENESEE BREWING CO., INC.
GENESEE BEER
ROCHESTER, N.Y.
CCC
ROUND 12"
MID-1950s

GENESEE BREWING CO., INC.
GENESEE CREAM ALE
ROCHESTER, N.Y.
MFG: ?
ROUND 12"
1960s

GENESEE BREWING CO., INC.
GENESEE BEER
ROCHESTER, N.Y.
CCC
ROUND 12"
LATE 1960s

GENESEE BREWING CO., INC.
GENESEE CREAM ALE
ROCHESTER, N.Y.
MFG: ?
ROUND 12"
1970s

GENEVA BREWING CO.
GENEVA HOME BREW
GENEVA, N.Y.
KAUFMANN AND STRAUSS
ROUND 12"
1910 - 1915

GERMAN AMERICAN BREWING CO.
MALTOSIA DARK AND PALE BEERS
BUFFALO, N.Y.
AMERICAN ART WORKS
SQUARE 14" x 14"
1909 - 1920

THE GREATER NEW YORK BREWERY, INC.
BARTEL'S BEER
NEW YORK, N.Y.
MFG: ?
ROUND 12"
1940 - 1942

GLOBE BREWING COMPANY, INC.
ROYAL STYLE ALE, DICTATOR BEER
UTICA, N.Y.
N.Y. IMPORTING CO.
ROUND 13" PIE
1933 - 1937

GREENWAY BREWERY CO., INC.
GREENWAY'S ALE AND LAGER
SYRACUSE, N.Y.
ELECTRO - CHEMICAL ENGRAVING CO., INC.
ROUND 12"
1933 - 1939

HABERLE BREWING CO.
CONGRESS BEER
SYRACUSE, N.Y.
HAEUSERMANN LITHO CO.
ROUND 12"
PRE - PRO

GULF BREWING COMPANY
GULF ALES
UTICA, N.Y.
HAEUSERMANN LITHO
ROUND 12"
1905 - 1920

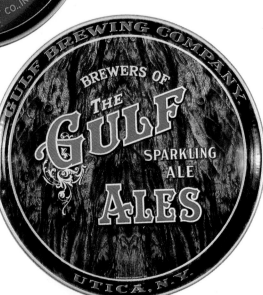

HABERLE BREWING CO.
CONGRESS BEER
SYRACUSE, N.Y.
AMERICAN ART WORKS
RECT. 11" x 14"
PRE - PRO

HABERLE BREWING CO.
CONGRESS BEER
SYRACUSE, N.Y.
HAEUSERMANN LITHO CO.
OVAL 8" x 11 1/2 "
1905 - 1919

HABERLE - CRYSTAL BOTTLING DEPT.
CONGO
SYRACUSE, N.Y.
SHONK WORKS - AMER. CAN CO (HAYWOOD, ILL)
ROUND 12"
1920 - 1933 (Prohibition era NEAR BEER)

HABERLE BREWING CO.
CONGRESS BEER
SYRACUSE, N.Y.
CHAS. W. SHONK
ROUND 12"
PRE - PRO

HABERLE CONGRESS BREWING CO., INC.
CONGRESS BEER
SYRACUSE, N.Y.
ELECTRO - CHEMICAL ENGRAVING CO., INC.
ROUND 12"
1936

HABERLE CONGRESS BREWING CO., INC.
CONGRESS BEER
SYRACUSE, N.Y.
N.Y. IMPORTING CO.
ROUND 13" PIE
1930s

HABERLE CONGRESS BREWING CO., INC.
DERBY CREAM ALE, CONGRESS BEER, BLACK BASS ALE
SYRACUSE, N.Y.
ELECTRO - CHEMICAL ENGRAVING CO., INC.
ROUND 12"
EARLY 1940s

HABERLE CONGRESS BREWING CO., INC.
BLACK RIVER ALE
SYRACUSE, N.Y.
ELECTRO - CHEMICAL ENGRAVING CO., INC.
ROUND 12"
1936

HEDRICK BREWING CO., INC.
HEDRICK ALE AND LAGER
ALBANY, N.Y.
MFG: ?
ROUND 13"
1940s

HABERLE CONGRESS BREWING CO., INC.
HABERLE CONGRESS LAGER, BLACK RIVER ALE,
LIGHT ALE
SYRACUSE, N.Y.
A.C. CO.
ROUND 12"
1940s

HEDRICK BREWING CO., INC.
HEDRICK ALE AND LAGER
ALBANY, N.Y.
AMERICAN CAN CO.
RECT. 11" x 14"
1930s

HEDRICK BREWING CO., INC.
HEDRICK BEER AND CREAM ALE
ALBANY, N.Y.
MFG: ?
ROUND 12"
LATE 1950s

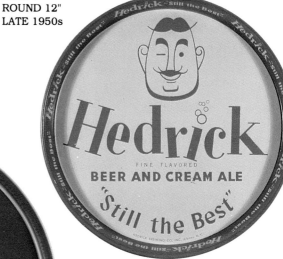

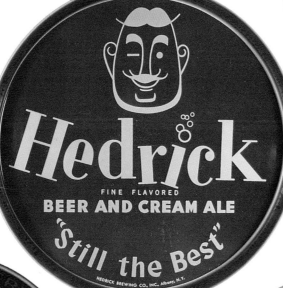

HEDRICK BREWING CO., INC.
HEDRICK BEER AND CREAM ALE
ALBANY, N.Y.
MFG: ?
ROUND 12"
LATE 1950s

HINCKEL BREWING CO.
HINCKEL'S LAGER - ALES
ALBANY, BOSTON, MANCHESTER
CHAS. W. SHONK
ROUND 12" PIE
1903 - 1919

HITTLEMAN GOLDENROD BREWERY CO.
GOLDENROD BEER, PORTER, ALE
BROOKLYN, N.Y.
AMERICAN CAN CO.
OVAL 14" x 17"
1934 - 1937

HORNELL BREWING CO., INC.
HORNELL LAGER BEER - CRYSTAL ALE
HORNELL, N.Y.
MFG: ?
ROUND 12"
1940s

F. HOLLENDER & CO.
IMPORTERS & EXPORTERS
NEW YORK, N.Y.
MFG: ?
OVAL 14" x 17"
PRE - PRO
*note - Hollander was not a brewery but a distributor

HORTON PILSNER BREWING CO., INC.
HORTON BEER & ALE
NEW YORK, N.Y.
AMERICAN CAN CO.
ROUND 12"
1933 - 1941

HORNELL BREWING CO., INC.
SCHWARTZENBRAU - K.D.K. CREAM ALE
HORNELL, N.Y.
AMERICAN ART WORKS, INC.
ROUND 12"
1940

HORTON PILSNER BREWING CO., INC.
HORTON BEER
NEW YORK, N.Y.
A.C. CO.
ROUND 13"
1933 - 1941

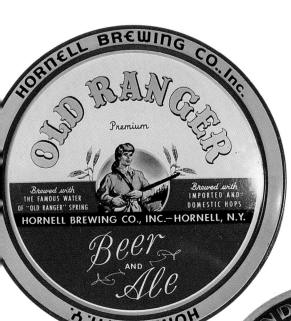

INDEPENDENT BREWING CO.
OLD TIMES LAGER, XXXX ALES
AUBURN, N.Y.
CHAS. W. SHONK
ROUND 12" PIE
1902 - 1916

HORNELL BREWING CO., INC.
OLD RANGER BEER AND ALE
HORNELL, N.Y.
H. D. BEACH CO.
ROUND 12"
EARLY 1940s

J. CHR. G. HUPFEL BRG. CO.
HUPFEL WIENER BEER
NEW YORK, N.Y.
KAUFMANN & STRUASS
ROUND 12"
I believe this tray is PRE - PRO

IROQUOIS BREWERY
IROQUOIS BEER
BUFFALO, N.Y.
CHAS. W. SHONK
"MINNE-HA-HA"
ROUND 12" PIE
PRE - PRO

OTTO HUBER BREWERY
PALE ALE, LAGER BEER
BROOKLYN, N.Y.
MFG: ?
"A CHIP OF THE OLD BLOCK"
OVAL 14" x 17"

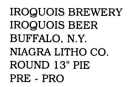

IROQUOIS BREWERY
IROQUOIS BEER
BUFFALO, N.Y.
NIAGRA LITHO CO.
ROUND 13" PIE
PRE - PRO

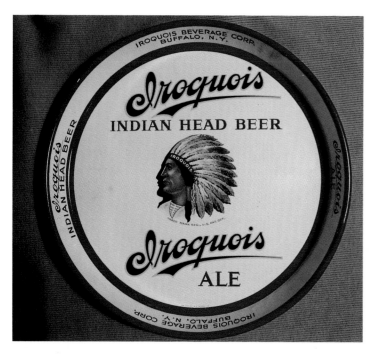

IROQUOIS BEVERAGE CORP.
IROQUOIS INDIAN HEAD BEER, ALE
BUFFALO, N.Y.
AMERICAN CAN CO.
ROUND 12"
LATE 1940s, 1950s

INTERNATIONAL BREWERIES, INC.
IROQUOIS INDIAN HEAD BEER
BUFFALO, N.Y.
CCC
ROUND 13"
LATE 1950s

IROQUOIS BEVERAGE CORP.
IROQUOIS INDIAN HEAD BEER & ALE
BUFFALO, N.Y.
AMERICAN CAN CO.
ROUND 13"
1950s

JETTER BREWING CO.
JETTER LAGER - SPECIAL BREW, WEINER & PILSNER
NEW YORK, N.Y.
THE MEEK CO.
ROUND 13"
1904 - 1909

INTERNATIONAL BREWERIES, INC.
IROQUOIS INDIAN HEAD BEER
BUFFALO, N.Y.
CANCO
ROUND 13"
1960s

KIPS BAY BREWING CO., INC.
KIPS BAY ALES, PORTER, EXTRA BEERS
NEW YORK, N.Y.
ELECTRO - CHEMICAL ENGRAVING CO., INC.
ROUND 13"
MID-1930s

GERHARD LANG BREWERY
LANG
BUFFALO, N.Y.
HAEUSERMANN LITHO
RECT. 11" x 14"
PRE - PRO

FRED KOCH BREWERY
KOCH'S BEER
DUNKIRK, N.Y.
MFG: ?
ROUND 13" PIE (also in 12" straight sided)
1940s

KUHN'S BEER, INC.
KUHN'S BEER
JAMESTOWN, N.Y.
F. J. OFFERMANN ART WORKS, INC.
ROUND 12"
1933 - 1937

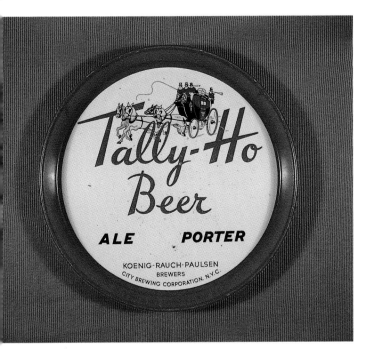

KOENIG-RAUCH-PAULSEN BREWERS
CITY BREWING CORPORATION
TALLY-HO BEER - ALE - PORTER
NEW YORK, N.Y.
AMERICAN CAN CO.
ROUND 12"
MID-1930s

GERHARD LANG BREWERY
HORSE HEAD BEER AND ALE
BUFFALO, N.Y.
A. A. W.
ROUND 12"
1940s

GERHARD LANG BOTTLING WORKS
LANG'S SOFT DRINKS
BUFFALO, N.Y.
AMERICAN ART WORKS
RECT. 11" x 14"
Prohibition ERA

GERHARD LANG BREWERY
OLD GERMAN BEER
BUFFALO, N.Y.
ELECTRO - CHEMICAL ENGRAVING CO., INC.
ROUND 12"
1930s

THE LAURER BEVERAGE CORP.
LAURER'S PILSNER BEER
BINGHAMTON, N.Y.
N.Y. IMPORTING CO.
ROUND 13" PIE
1933 - 1936

LIEBMANN BREWERIES, INC.
RHEINGOLD BEER & ALE
NEW YORK, N.Y.
OWENS - ILLINOIS CAN COMPANY
ROUND 13"
EARLY 1950s

LIEBMAN BREWERIES, INC.
RHEINGOLD SCOTCH ALE - BEER
BROOKLYN, N.Y.
AMERICAN CAN CO.
ARTIST: O. SOGLOW
ROUND 12"
1933 (Copyright date)

ROWLIEBMANN BREWERIES INC.
RHEINGOLD EXTRA DRY LAGER BEER
NEW YORK, N.Y. and from 1950 - 1964
also ORANGE, N.J.
AMERICAN (CANCO) or CONTINENTAL (CCC)
ROUND - usually 12" but also 13"
* These 4 variations are representative of the many similar RHEINGOLD trays of this design. This design has my vote as the most common of all metal beer trays. Picturing each design modification would require more pages than are devoted to Rhode Island trays.
ROUND 13"
1940s

LIEBMAN BREWERIES, INC.
RHEINGOLD SCOTCH ALE - BEER
BROOKLYN, N.Y.
AMERICAN CAN CO.
ARTIST: O. SOGLOW
ROUND 12"
1933 (Copyright date)

THE LION BREWERY
BUFFALO, N.Y.
CHAS. W. SHONK
ROUND 12" PIE
1892 - 1904

LINDEN BREWERY, INC.
LINDEN BEER, ALE - PORTER
LINDENHURST, LONG ISLAND, N.Y.
ELECTRO - CHEMICAL ENGRAVING CO., INC.
ROUND 12"
1938

LION BREWERY OF N.Y. CITY
LION BEER
NEW YORK, N.Y.
AMERICAN CAN CO.
OVAL 12 1/2" x 15
MID - LATE 1930s

LINDEN BREWERY, INC.
LINDEN BEER - ALE
LINDENHURST, N.Y.
OWENS - ILLINOIS CAN COMPANY
ROUND 13"
1940s

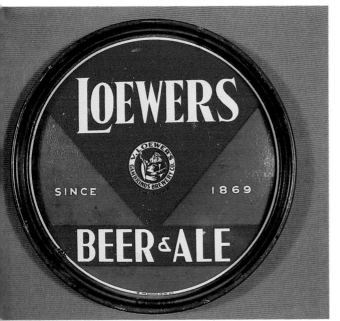

V. LOEWERS GAMBRINUS BREWERY CO.
LOEWERS BEER AND ALE
NEW YORK, N.Y.
THE BURDICK CO.
ROUND 12"
1933 - 1943

CHAS. LUTZ & BRO.
TEUTONIC DARK BEER, RHEINGOLD PALE BEER
BROOKLYN, N.Y.
AMERICAN ART WORKS
"PROSIT"
RECT 11" x 14"
CO. 1910

MONROE BREWING CO.
MONROE BEER
ROCHESTER, N.Y.
CHAS. W. SHONK
ROUND 12"
PRE - PRO

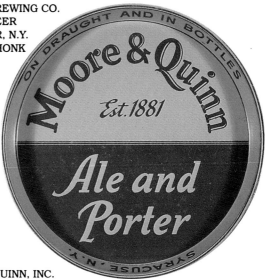

MOORE & QUINN, INC.
MOORE AND QUINN
SYRACUSE, N.Y.
UNIVERSAL TRAY AND SIGN CO.
ROUND 12"
1915 - 1920

MOORE & QUINN, INC.
DIAMOND ALE
SYRACUSE, N.Y.
ELECTRO - CHEMICAL ENGRAVING CO., INC.
ROUND 13"
1938

NECTAR BREWING CORP.
NECTAR LAGER BEER - SPARKLING ALE - XXXX PORTER
ELMIRA, N.Y.
ELECTRO - CHEMICAL ENGRAVING CO., INC.
ROUND 12"
1933 - 1939

NIAGARA FALLS BREWING CO.
SPRAY BEER
NIAGARA FALLS, N.Y.
HAEUSERMANN LITHO CO.
"THE TOP NOTCHER OF THEM ALL"
RECT 11" x 14"
PRE - PRO

NORTH AMERICAN BREWING CO.
PARAMOUNT SUPER BEER AND ALE
BROOKLYN, N.Y.
AMERICAN CAN CO.
ROUND 12"
1933 - 1946

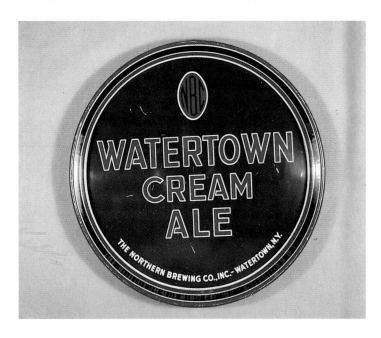

OLD DUTCH BREWERS, INC.
OLD DUTCH LAGER BEER, ALE & PORTER
BROOKLYN, N.Y.
AMERICAN CAN CO.
ROUND 12"
1934 - 1948

THE NORTHERN BREWING CO., INC.
WATERTOWN CREAM ALE
WATERTOWN, N.Y.
ELECTRO - CHEMICAL ENGRAVING CO., INC.
ROUND 12"
1933 - 1943

OLD DUTCH BREWERS, INC.
DUTCH TAVERN LAGER BEER
BROOKLYN, N.Y.
NEW YORK IMPORTING CO.
ROUND 13" PIE
MID TO LATE 1930s

ORANGE COUNTY BREWERY, INC.
GOLDEN FOAM BEER
MIDDLETOWN, N.Y.
ELECTRO - CHEMICAL ENGRAVING CO., INC.
ROUND 12"
1933 - 1934

ONEIDA BREWING CO.
ONEIDA ALES AND PORTER
UTICA, N.Y.
CHAS. W. SHONK
"RODERICK"
ROUND 12" PIE
CO. 1904

PHOENIX BREWERY CORP.
PHOENIX BEER
BUFFALO, N.Y.
F. J. OFFERMANN ART WORKS, INC.
ROUND 12"
1930s

ONEIDA BREWING CO., INC.
ONEIDA SPARKLING CREAM ALE
TROY, N.Y.
ELECTRO - CHEMICAL ENGRAVING CO., INC.
ROUND 12"
1933 - 1942

PHOENIX BREWERY CORP.
PHOENIX BEER - MOFFATS ALE
BUFFALO, N.Y.
ELECTRO - CHEMICAL ENGRAVING CO., INC.
ROUND 13"
1936

PHOENIX BREWERY CORP.
PHOENIX ALE - BEER
BUFFALO, N.Y.
A.C. CO.
ROUND 13"
1940s

PHOENIX BREWERY
BUFFALO, N.Y.
CHAS. W. SHONK
OVAL 14" x 19"
PRE - PRO

PHOENIX BREWERY CORP.
PHOENIX BEER - MOFFATS ALE
BUFFALO, N.Y.
AMERICAN CAN CO.
ROUND 13"
1940s

ROWPIEL BROS., INC.
PIELS LIGHT BEER
NEW YORK, N.Y.
MFG: A.C. CO., CCC, CANCO (left to right)
ROUND 12"
ALL 1950s, EARLY 1960s

PIEL BROS., INC.
PIELS
BROOKLYN, N.Y.
ELECTRO - CHEMICAL ENGRAVING CO., INC.
ROUND 14" PIE
C. 1934

PIEL BROS., INC.
PIEL'S LAGER BEER
BROOKLYN, N.Y.
ELECTRO - CHEMICAL ENGRAVING CO., INC.
ROUND 12"
1930s
* The two Piel's Lager Beer trays are identical except that the elves on the left side tray are drawn like cartoon caricatures, whereas on the right they appear to be carefully painted, with an almost airbrushed realism to them. (Plus they each lost about 50 lbs. of beer belly)

PIEL BROS., INC.
PIEL'S LAGER BEER
BROOKLYN, N.Y.
ELECTRO - CHEMICAL ENGRAVING CO., INC.
ROUND 12"
1930s

PIEL BROS., INC.
PIELS BEER
BROOKLYN, N.Y.
THE BURDICK CO.
ROUND 13"
1930s

PIEL BROS., INC.
PIELS BEER
BROOKLYN, N.Y.
CCC
ROUND 12"
C. 1963

PIEL BROS., INC.
PIELS
BROOKLYN, N.Y.
CCC
ROUND 12"
C. 1961

PIEL BROS., INC.
PIELS BEER
BROOKLYN, N.Y.
CANCO
ROUND 13"
C. 1957

ROWPIEL BROS., INC.
PIELS BEER
BROOKLYN, N.Y.
CANCO
ROUND 13"
C. 1957

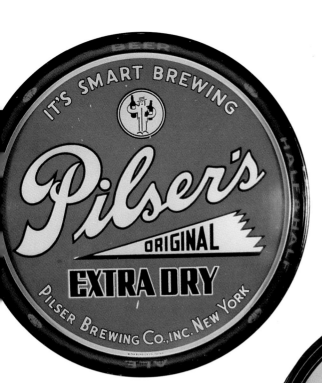

RECORD BREWING CO.
RECORD LAGER, ALES & PORTER
ELMIRA, N.Y.
HAEUSERMANN LITHO CO.
ROUND 12"
1907 - 1919

PILSNER BREWING COMPANY, INC.
PILSNER'S ORIGINAL EXTRA DRY
NEW YORK, N.Y.
THE BURDICK CO.
ROUND 12"
1930s

QUANDT BREWING CO.
QUANDT LAGER - ALES
TROY, N.Y.
CHAS. W. SHONK
ROUND 12"
PRE - PRO

RECORD BREWING CO.
GOLD STANDARD LAGER, CREAM ALE
ELMIRA, N.Y.
HAEUSERMANN LITHO CO.
ROUND 12"
1907 - 1919

GEO. RINGLER & CO.
ROYAL PALE
NEW YORK, N.Y.
THE H. D. BEACH CO.
ROUND 12"
PRE - PRO

ROCHESTER BREWING CO., INC.
OLD TOPPER ALE
ROCHESTER, N.Y.
ELECTRO - CHEMICAL ENGRAVING CO., INC.
ROUND 12"
1937

GEO. RINGLER & CO.
EXTRA PILSNER AND REAL GERMAN BEERS
NEW YORK, N.Y.
THE H. D. BEACH CO. (in slanted italic lettering)
OVAL 14" X 17"
PRE - PRO

ROCHESTER BREWING CO., INC.
GOLDEN OLD TOPPER ALE & BEER
ROCHESTER, N.Y.
AMERICAN CAN CO.
ROUND 12"
EARLY 1950s

ROME BREWERY, INC. (successors to EVANS & GIEHL, INC.)
ROME ALE
ROME, N.Y.
NEW YORK IMPORTING CO.
ROUND 13" PIE
1933 - 1935

RUBSAM AND HORRMANN BREWING CO.
CROWN LAGER BEER
NEW YORK, N.Y.
AMERICAN CAN CO.
ROUND 13"
1940s

RUBSAM & HORRMANN BR'G. CO.
R & H BEER
STATEN ISLAND, N.Y.
MFG: ?
ROUND 13" PIE
1930s

RUBSAM AND HORRMANN BR'G. CO.
R & H BEER - ALE
STATEN ISLAND, N.Y.
A.C. CO.
ROUND 13"
LATE 1940s

RUBSAM AND HORRMANN BREWING CO.
R & H LIGHT BEER
NEW YORK CITY, N.Y.
MFG: ?
ROUND 12"
LATE 1940s, EARLY 1950s

JACOB RUPPERT
JACOB RUPPERT BEER
NEW YORK, N.Y.
ELECTRO - CHEMICAL
ENGRAVING CO., INC.
ROUND 13" PIE
1937

JACOB RUPPERT
JACOB RUPPERT'S LAGER BEER
NEW YORK, N.Y.
AMERICAN ART WORKS
ARTIST: E. H. KIEFER (This artist also created the
Huber tray artwork for the tray entitled " A CHIP OF THE OLD
BLOCK")
SQUARE 14" x 14"
1909 - 1919

JACOB RUPPERT, INC.
JACOB RUPPERT BEER - ALE
NEW YORK, N.Y.
ELECTRO - CHEMICAL ENGRAVING CO., INC.
OVAL 10 1/2 " x 14 1/2"
1938
A round 12" version of this tray, also by Electro - Chemical, 1939, exists

JACOB RUPPERT
ROSE BUD BEER
NEW YORK, N.Y.
THE MEEK CO.
OVAL 14" x 17"
C. 1905

JACOB RUPPERT
KNICKERBOCKER BEER
NEW YORK, N.Y.
MFG: (2248-100M358 - JACOB RUPPERT)
N.Y.C. (printed on outer top side wall under rim)
ROUND 12"
LATE 1950s, EARLY 1960s

JACOB RUPPERT
ROSE BUD BEER
NEW YORK, N.Y.
H. D. BEACH
OVAL 14" x 17"
1905 - 1910

JACOB RUPPERT
KNICKERBOCKER BEER
NEW YORK, N.Y.
CANCO
ROUND 13"
1950s

JACOB RUPPERT
KNICKERBOCKER BEER
NEW YORK, N.Y.
CCC
ROUND 12"
LATE 1950s, EARLY 1960s

JACOB RUPPERT
KNICKERBOCKER BEER
NEW YORK, N.Y.
CANCO
ROUND 13"
1950s

THE F. & M. SCHAEFFER BREWING CO.
SCHAEFFER BEER
BROOKLYN, N.Y.
A.C. CO.
ROUND 13"
MID-1930s

THOS. RYAN'S CONSUMERS BREWING CO.
ONONDAGA LAGER - SPARKLING ALES
SYRACUSE, N.Y.
CHAS. W. SHONK
ROUND 12"
PRE-1919

F. & M. SCHAEFFER BREWING CO.
SCHAEFFER BEER
BROOKLYN, N.Y. and ALBANY, N.Y.
CANCO
ROUND 12"
LATE 1950s

THE F. & M. SCHAEFFER BREWING CO.
WIENER BEER, SPECIAL DARK & LAGER
NEW YORK, N.Y.
PALM, FECHTLER & CO.
OVAL 12" x 15"
PRE - PRO

F. & M. SCHAEFFER BREWING CO.
SCHAEFFER BEER
BROOKLYN, N.Y.
AMERICAN CAN CO.
ROUND 12"
1950s

SCHREIBER BREWING CO., INC.
SCHREIBER'S MANRU BEER - ALE
BUFFALO, N.Y.
ELECTRO - CHEMICAL ENGRAVING CO., INC.
ROUND 13"
C. 1937

F. & M. SCHAEFFER BREWING CO.
SCHAEFFER BEER
NEW YORK and ALBANY, N.Y.
CCC
ROUND 12"
LATE 1970s

WILLIAM SIMON BREWERY
SIMON PURE BEER, OLD ABBEY ALE
BUFFALO, N.Y.
ELECTRO - CHEMICAL ENGRAVING CO., INC.
ROUND 13"
1936

SCHREIBER BREWING CO., INC.
MANRU KLOSTER
BUFFALO, N.Y.
AMERICAN CAN CO.
ROUND 12"
LATE 1930s, 1940s

WM. SIMON BREWERY
SIMON PURE
BUFFALO, N.Y.
HAEUSERMANN LITHO
RECT 11" x 14"
1905 - 1920

FRANK X. SCHWAB CO.
GENEVA ALE & PORTER
BUFFALO, N.Y.
HAEUSERMANN LITHO
RECT. 11" x 14"
PRE - PRO

*NOTE: The Francis X. Schwab Brewing Corp. 1933-1933 is the only Schwab Brewery listed in American Breweries (Bull et. al.)

Since Haeusermann Litho closed in 1921, it seems Schwab was originally a distributor/wholesaler.

WILLIAM SIMON BREWERY
SIMON PURE ALE - BEER
BUFFALO, N.Y.
A.C. CO.
ROUND 12"
1940s

WILLIAM SIMON BREWERY
SIMON PURE BEER, OLD ABBEY ALE
BUFFALO, N.Y.
CANCO
ROUND 13"
1960s

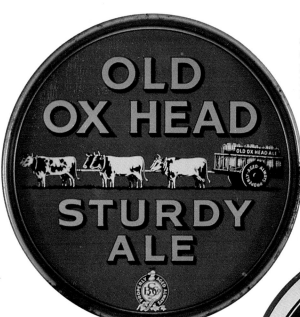

STANDARD BREWING CO., INC.
OLD OX HEAD STURDY ALE
ROCHESTER, N.Y.
ELECTRO - CHEMICAL ENGRAVING CO., INC.
ROUND 12"
1936

STANDARD BREWING CO., INC.
OX CART BEER, STANDARD DRY A
ROCHESTER, N.Y.
A.C. CO.
ROUND 12"
LATE 1940s, EARLY 1950s

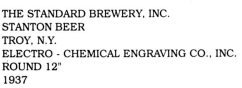

STANDARD BREWING CO., INC.
STANDARD ALE
ROCHESTER, N.Y.
MFG: ?
ROUND 12"
1940s

THE STANDARD BREWERY, INC.
STANTON BEER
TROY, N.Y.
ELECTRO - CHEMICAL ENGRAVING CO., INC.
ROUND 12"
1937

THE STANTON BREWERY, INC.
STANTON BEER
TROY, N.Y.
ELECTRO - CHEMICAL ENGRAVING CO., INC.
ROUND 12"
1935

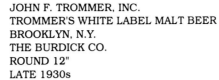

JOHN F. TROMMER, INC.
TROMMER'S WHITE LABEL MALT BEER
BROOKLYN, N.Y.
THE BURDICK CO.
ROUND 12"
LATE 1930s

GEO. F. STEIN BREWERY, INC.
STEINS BEER
BUFFALO, N.Y.
CANCO
ROUND 13"
1950s

SYRACUSE BREWERY, INC.
DICKEN'S ALE
SYRACUSE, N.Y.
AMERICAN CAN CO.
ROUND 12"
1934 - 1937

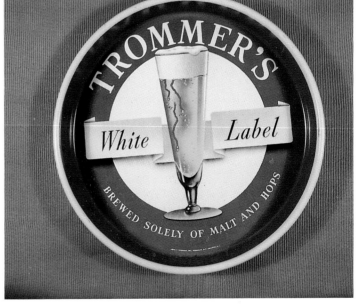

JOHN F. TROMMERS, INC.
TROMMER'S MALT BEERS
BROOKLYN, N.Y.
MASON CAN CO.
ROUND 13"
1940s

JOHN F. TROMMERS, INC.
TROMMER'S WHITE LABEL BEER
BROOKLYN, N.Y., ORANGE, N.J.
CCC
ROUND 13"
LATE 1940s

GEORGE ZETT BREWERY
ZETT'S BAVARIAN BEER
SYRACUSE, N.Y.
AMERICAN ART WORKS
SQUARE 14" x 14"
1909 - 1919

WELZ & ZERWECK BREWERY
HIGH GROUND BREWERY
GAMBRINUS BRAU
BROOKLYN, N.Y.
HAEUSERMANN LITHO
OVAL 14" x 17"
1905 - 1920

WEST END BREWING CO.
UTICA BEER
UTICA, N.Y.
MFG: ?
"AULD LANG SYNE" STOCK #107
RECT. 11" x 14"
C. 1910

WEST END BREWING CO.
UTICA CLUB BEER
UTICA, N.Y.
CCC
ROUND 13"
1960s

WEST END BREWING CO.
WUERZBURGER, PILSNER, ALE, STOCK AND PORTER
UTICA, N.Y.
KAUFMANN & STRAUSS
ROUND 13" PIE
PRE - PRO

WEST END BREWING CO.
UTICA CLUB
UTICA, N.Y.
NEW YORK IMPORTING CO.
ROUND 13" PIE
MID - LATE 1940s

GEORGE ZETT BREWING CO., INC.
PAR-EX BEER
SYRACUSE, N.Y.
MFG: ?
ROUND 12"
1933 - 1934
(possibly
PRE - PRO)

YONKERS COLONIAL CORP.
YONKERS ALE - BEER
YONKERS, N.Y.
ELECTRO - CHEMICAL ENGRAVING CO., INC.
ROUND 12"
1937

GEORGE ZETT BREWERY
PAR-EX
SYRACUSE, N.Y.
KAUFMANN & STRUASS
"PURITY"
ROUND 12"
PRE - PRO

New Hampshire

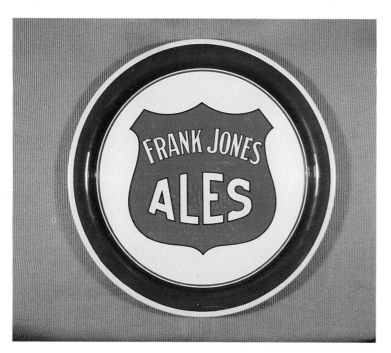

FRANK JONES BREWING CO.
FRANK JONES ALES
PORTSMOUTH, N.H.
KAUFMANN & STRAUSS
ROUND 13" PIE
PRE - PRO

FRANK JONES BREWING CO., INC.
FRANK JONES ALES
PORTSMOUTH, N.H.
MFG: ?
ROUND 13" PIE
1930s

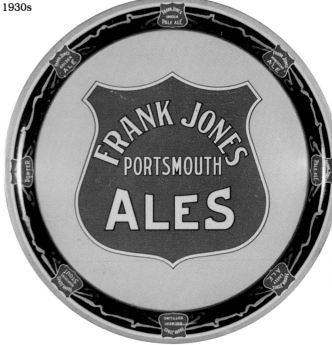

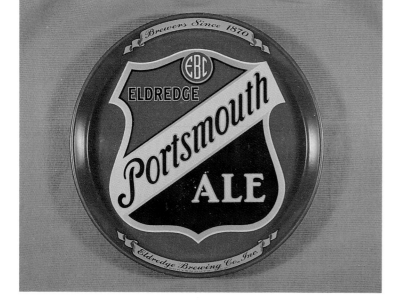

ELDRIDGE BREWING CO.
PORTSMOUTH ALE
PORTSMOUTH, N.H.
AMERICAN CAN CO.
ROUND 12"
EARLY 1940s

TRUE W. JONES BREWING CO.
GRANITE STATE & CUMAX ALES, JONES' LAGER BEER
MANCHESTER, N.H.
AMERICAN ART WORKS
ROUND 13"
1909 - 1918

FRANK JONES BREWING CO., INC.
FRANK JONES ALES
PORTSMOUTH, N.H.
MFG: ?
ROUND 12
1940s

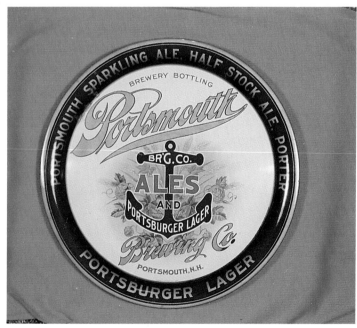

PORTSMOUTH BRG. CO.
PORTSMOUTH ALES - PORTSBURGER LAGER
PORTSMOUTH, N.H.
KAUFMANN & STRAUSS
ROUND 13" PIE
PRE - PRO

Connecticut

ROPKINS & CO.
ROPKINS BITTER ALE, LIGHT DINNER ALES
ROPKINS SUPERIOR PORTER
HARTFORD, CT.
CHAS. W. SHONK
OVAL 14" x 17"
PRE - PRO

THE HARTMANN BREWING CO.
BRIDGEPORT, CT.
THE MEEK AND BEACH CO.
STOCK #47, THEODORE ROOSEVELT
OVAL 14" x 17"
C. 1903

THE HUBERT FISCHER BREWERY
BOHEMIAN LAGER - BRILLIANT ALE
HARTFORD. CT.
THE MEEK CO.
OVAL 16" x 19"
1901 - 1909

AETNA BREWING COMPANY
AETNA SPECIAL DINNER ALE
HARTFORD, CT.
UNIVERSAL TRAY AND SIGN CO.
ROUND 12"
1933 - 1939

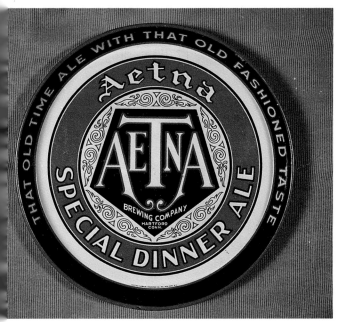

AETNA BREWING COMPANY
AETNA SPECIAL DINNER ALE
HARTFORD, CT.
ELECTRO - CHEMICAL ENGRAVING CO., INC
ROUND 12"
1933 - 1939

AETNA BREWING CO.
HARTFORD, CT.
MFG: ?
PORCELAIN
OVAL 14" x 17"
1900 - 1910

BRIDGEPORT BREWING COMPANY
PIONEER ALES AND LAGER
BRIDGEPORT, CT.
UNIVERSAL TRAY AND SIGN CO.
ROUND 12"
1934 - 1941

CREMO BREWING CO., INC.
CREMO ALE - LAGER
NEW BRITAIN, CT.
MFG: ?
ROUND 13"
1930s

CREMO BREWING COMPANY INC.
CREMO ALE AND LAGER
NEW BRITAIN, CT.
UNIVERSAL TRAY AND SIGN CO.
ROUND 12"
1930s

THE ELM CITY BREWING CO.
HAGEARTY'S BREW
NEW HAVEN, CT.
AMERICAN CAN CO.
OVAL 13" x 16"
1933 - 1936

CREMO BREWING CO.
CREMO ALE AND LAGER
NEW BRITAIN, CO.
UNIVERSAL TRAY AND SIGN CO.
ROUND 12"
1930s

CREMO BREWING CO.
CREMO SPARKLING ALE
NEW BRITAIN, CT.
MFG: ?
ROUND 13"
1940s

CREMO BREWING CO.
CREMO SPARKLING ALE
NEW BRITAIN, CT.
MFG: ?
ROUND 13"
1940s

HUBERT FISCHER BREWERY
LAGER BEER
HARTFORD, CT.
CHROMO. ENG. MAKERS, SAML. BUCKLEY & CO.
ROUND 12" PIE (porcelain)
PRE - PRO

THE HUBERT FISCHER BREWERY
COBURGER BEER, SPECIAL BREW, BRILLIANT ALE
HARTFORD, CT.
KAUFMANN & STRAUSS
RECT. 11" x 14"
PRE - PRO

HUBERT FISCHER
HARTFORD, CT.
F. E. MARSLAND
ROUND 12" PIE - ETCHED PLATED TIN
1886 - 1894

HUBERT FISCHER BREWERY
LAGER BEER
HARTFORD, CT.
THE INTERNATIONAL AD. CO.
ROUND 12" PIE (porcelain)
PRE - PRO

THE HUBERT FISCHER BREWERY
LAGER, ALES & PORTER
HARTFORD, CT.
THE BALTIMORE ENAMEL AND NOVELTY CO.
ROUND 12" PIE (porcelain)
PRE - PRO

HARTMANN BREWING CO.
LAGER, ALES & PORTER
BRIDGEPORT, CT.
THE BALTIMORE ENAMEL AND NOVELTY CO.
OVAL 13" x 17" (porcelain)
1897 - 1911

THE HULL BREWING CO.
HULL'S CREAM ALE
NEW HAVEN, CT.
ELECTRO - CHEMICAL ENGRAVING CO., INC.
ROUND 12"
LATE 1930s

THE HULL BREWING CO.
HULL'S CREAM ALE - EXPORT BEER
NEW HAVEN, CT.
A.C. CO.
ROUND 12"
1940s

THE HULL BREWING CO.
HULL'S ALE - LAGER
NEW HAVEN, CT.
MFG: ?
ROUND 13" PIE
LATE 1930s, 1940s

THE HULL BREWING CO.
HULL'S CREAM ALE - LIGHT BEER
NEW HAVEN, CT.
A.C. CO.
ROUND 12" PIE
1940s

THE LARGAY BREWING CO., INC.
RED FOX ALE
WATERBURY, CT.
THE H. D. BEACH CO.
ROUND 12"
LATE 1930s

THE LARGAY BREWING CO., INC.
RED FOX ALE - BEER
WATERBURY, CT.
AMERICAN ART WORKS
ROUND 12"
1940

*Note - The two Red Fox Ale trays are commonly referred to as "SPINNER" trays. They have an indentation punched into the center of the back side so that they can be balanced on a pencil or pen point. Notice that the six small foxes on the tray rim are shown in two alternating positions. When the trays are balanced and spun rapidly, the observer with a vivid imagination and a number of emptys in front of him may experience the visual illusion that the fox is running.

THE HULL BREWING CO.
HULL'S LAGER BEER - CREAM ALE
NEW HAVEN, CT.
UNIVERSAL TRAY AND SIGN CO.
ROUND 12"
1930s

THE HULL BREWING CO.
HULL'S EXPORT BEER
NEW HAVEN, CT.
CCC
ROUND 13"
1960s

THE LARGAY BREWING CO., INC.
RED FOX ALE & LAGER
WATERBURY, CT.
ELECTRO - CHEMICAL ENGRAVING CO., INC.
ROUND 13"
MID-1930s

THE NEW ENGLAND BREWING COMPANY
ALES & LAGER
HARTFORD, CT.
THE METALLOGRAPH CORP.
ROUND 12"
1936 - 1943

THE LARGAY BREWING CO., INC.
WHITE CAP BEER
WATERBURY, CT.
AMERICAN ART WORKS
ROUND 12"
LATE 1930s, EARLY 1940s

THE NEW ENGLAND BREWING CO.
EXTRA CABINET LAGER, SUPERIOR FAMILY ALE
HARTFORD, CT.
HAEUSERMANN LITHO
OVAL 14" x 17"
1905 - 1915

NEW ENGLAND BREWING CO.
SUPERIOR ALES & LAGER
HARTFORD, CT.
BALTIMORE ENAMEL AND NOVELTY CO.
ROUND 12" PIE (porcelain)
1900 - 1910

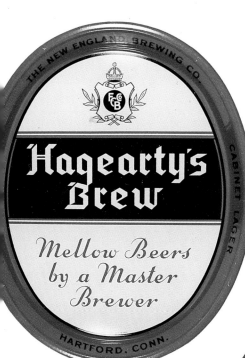

OLD ENGLAND BREWING CO., INC.
OLD ENGLAND ALE AND LAGER
DERBY, CT.
MASON CAN CO.
ROUND 13"
1933 - 1941

THE NEW ENGLAND BREWING CO.
HAGEARTY'S BREW
HARTFORD, CT.
AMERICAN CAN CO.
OVAL 13" x 16"
1936 - 1943

THE NEW ENGLAND BREWING CO.
NEBCO ALE AND LAGER
HARTFORD, CT.
MFG: ?
ROUND 13" PIE
1936

WATERBURY BREWING COMPANY
LEARY'S CLOCK ALE - LAGER
WATERBURY, CT.
MFG: ?
ROUND 13"
1934 - 1938

ROPKINS & CO.
ROPKINS LIGHT DINNER ALE
HARTFORD, CT.
AMERICAN ART WORKS
"YAMA YAMA" STOCK#115
RECT. 11" x 14"
PRE - PRO

WATERBURY BREWING COMPANY
CLOCK ALE - LAGER
WATERBURY, CT.
ELECTRO - CHEMICAL ENGRAVING CO., INC
ROUND 12"
1934 - 1938

THE WEHLE BREWING COMPANY
MULE HEAD ALE, OX HEAD BEER
WEST HAVEN, CT.
MFG: ?
ROUND 13" PIE
1930s

THE WEHLE BREWING COMPANY
MULE HEAD
WEST HAVEN, CT.
MFG: ?
ROUND 13" PIE
1930s

WEIBEL BREWING CO.
WEIBEL'S ALE & LAGER
NEW HAVEN, CT.
ACME SIGNS - DISPLAYS
ROUND 13"
1933 - 1936

THE WEHLE BREWING COMPANY
WEHLE ALE - BEER
WEST HAVEN, CT.
MFG: ?
ROUND 12"
1930s

WEIBEL'S BREWING CO.
WEIBEL'S GOLDEN ALE
NEW HAVEN, CT.
MFG: ?
ROUND 13" PIE
1933 - 1936

THE WEHLE BREWING COMPANY
MULE HEAD
WEST HAVEN, CT.
MFG: ?
ROUND 13" PIE
1930s

Pennsylvania

CASEY & KELLY
SCRANTON, PA.

E. ROBINSON'S SONS
SCRANTON, PA.

*Note - These four Pennsylvania examples are grouped together to illustrate a popular design used by numerous brewers which is commonly called a "STOCK" tray. The small brewer saved the expense of artwork by choosing a tray the manufacturer had previously commissioned and already had litho plates available for. The only difference distinguishing them is the information specifying brewery and brand identifications. These trays are all PRE-PRO 14" x 17" ovals made by THE H.D. BEACH CO.

PETER KRANTZ BREWERY
CARBONDALE, PA.

REICHARD & WEAVER
WILKES-BARRE, PA

A.B. CO. BREWERS
ROBIN HOOD ALE - WURZBURGER & BOHEMIAN BEERS
SCRANTON, PA.
ELECTRO - CHEMICAL ENGRAVING CO., INC.
OVAL 8" x 11 1/2"
1934 - 1937

ANDERTON BREWING CO.
NONPAREIL BEER, ANGLICE ALE, EXPORT, PORTER
BEAVER FALLS, PA.
MFG: ?
OVAL 14" x 17"
1891 - 1904

BARBEY'S, INC.
SUNSHINE BEER
READING, PA.
THE BURDICK CO.
ROUND 12"
EARLY 1940s

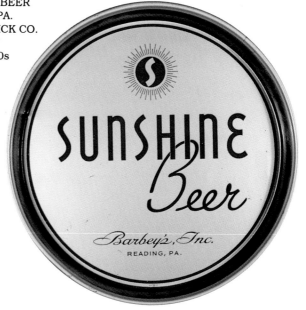

ARNHOLT - SCHAEFER BREWING CO.
ARNHOLT - SCHAEFER BEER
PHILADELPHIA, PA.
PASSAIC METAL WARE CO.
ROUND 12 3/4"
C. 1912

BARTELS BREWING CO.
BARTELS BEER
WILKES-BARRE, PA.
MFG: This tray is an example made during WWII. It is composed of the molded plastic-like synthetic known as bakelite. A handful of breweries distributed trays made of bakelite due to the tin shortage during the war years. This tray and a Stegmaier bakelite are oddities that I felt it was valuable to include. They are the only plastic the reader will find examples of here. They are ugly before their time and a grim omen of things to come. I decided to include these trays because, unlike today's plastic trays, they possess one redeeming virtue. Their creation was not the result of a cheapest-way-out option motivated by profit margin greed. Bakelite is symbolic of one small way private industry contributed to the war effort. Whenever a pitcher of Steg or Bartels ordered by patrons arrived at a tavern table atop a bakelite tray, civilian hearts should have swelled with patriotic pride. I suspect that even the beer tasted better, knowing that the round 12" quantity of metal composing the usual tray their beer arrived on was instead a small part of the military machine which eventually would defeat the Axis powers and preserve democracy. Cheers to Bartels and Stegmaier. God Bless America.

BARTELS BREWING CO.
$5,000.00 BEER, FINE ALES & PORTER
EDWARDSVILLE, PA.
KAUFMANN & STRAUSS
ROUND 13" PIE
1898 - 1906

BARTELS BREWING CO.
BARTELS BEER
WILKES-BARRE, PA.
CHAS. W. SHONK
ROUND 12"
1907 - 1919

BETH UHL BREWING CO.
TANNHAUSER BEER
BETHLEHEM, PA.
ELECTRO - CHEMICAL ENGRAVING CO., INC.
ROUND 13"
1936

JOHN F. BETZ AND SON, INC.
BETZ BEER
PHILADELPHIA, PA.
ELECTRO - CHEMICAL ENGRAVING CO., INC.
ROUND 12"
1933 - 1939

BINDER BREWING COMPANY
RENOVO, PA.
MFG: ?
ROUND 12"
1910 - 1920

BRACKENRIDGE BREWING CO., INC.
OLD ANCHOR BEER
BRACKENRIDGE, PA.
MFG: ?
ROUND 12"
1933 - 1940

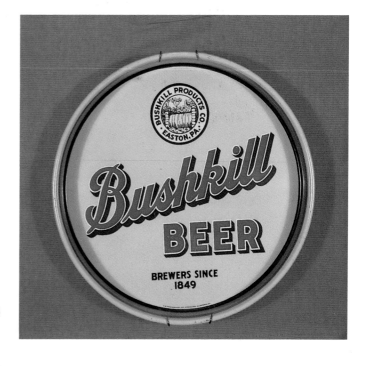

BUSHKILL PRODUCTS CO.
BUSHKILL LAGER BEER
EASTON, PA.
ACME
ROUND 13"
1933 - 1941

BUSHKILL PRODUCTS CO.
BUSHKILL BEER
EASTON, PA.
R. G. FENSTERMAKER ADV. SPECIALTIES
ROUND 12"
1933 - 1941

CASEY & KELLY BREWERY
PEERLESS BEER
SCRANTON, PA.
HAEUSERMANN LITHO
ROUND 13" PIE
1905 - 1919

CHESTER BREWERY, INC.
CHESTER BEER AND ALE
CHESTER, PA.
UNIVERSAL TRAY AND SIGN CO.
ROUND 12"
LATE 1930s

CHESTER BREWERY, INC.
CHESTER PILSNER BEER AND ALE
CHESTER, PA.
ELECTRO - CHEMICAL ENGRAVING CO., INC.
ROUND 12"
1935

CHESTER BREWERY, INC.
CHESTER PILSNER, ALE AND PORTER
CHESTER, PA.
UNIVERSAL TRAY AND SIGN CO.
ROUND 12"
MID - LATE 1930s

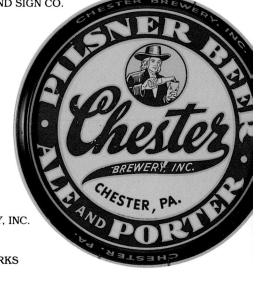

CHESTER BREWERY, INC.
CHESTER ALE
CHESTER, PA.
AMERICAN ART WORKS
ROUND 12"
1930s

COLUMBIA BREWING CO.
BEER, ALE, AND PORTER
SHENANDOAH, PA.
AMERICAN ART WORKS
"JOIN ME"
RECT 11" x 14"
PRE - PRO

COLUMBIA BREWING CO.
COLUMBIA BEER
SHENANDOAH, PA.
THE BURDICK CO.
ROUND 12"
LATE 1930s

COLUMBIA BREWING CO.
COLUMBIA 5 STAR BEER
SHENANDOAH, PA.
CCC
ROUND 12"
LATE 1940s, EARLY 1950s

CHESTER BREWERY, INC.
SILVER DIME SPECIAL BEER
CHESTER, PA.
MFG: ?
ROUND 12"
1930s

CHESTER BREWERY, INC.
SILVER DIME PREMIUM, CHESTER PILSNER
CHESTER, PA.
UNIVERSAL TRAY AND SIGN CO.
ROUND 12"
LATE 1930s

115

COLUMBIA BREWING CO.
COLUMBIA PREFERRED BEER
SHENANDOAH, PA.
CCC
ROUND 12"
1950s

COOPER BREWING CO.
COOPER'S OLD BOHEMIAN BEER
PHILADELPHIA, PA.
A.C. CO.
ROUND 12"
LATE 1930s

COOPER BREWERY
LIEBERT & OBERT
NAMAR PREMIUM BEER
PHILADELPHIA, PA.
MFG: ?
ROUND 12"
C. 1943

CRESSON SPRINGS BREWERY CO.
MEISTER BRAU - PRIMA BEER
CRESSON, PA.
KAUFMANN & STRAUSS
ROUND 13" PIE
1915 - 1920

DAEUFFER LIEBERMAN BREWERY
DAEUFER'S BEER
ALLENTOWN, PA.
AMERICAN CAN CO.
OVAL 12 1/2" x 15 1/2"
LATE 1930s

DUBOIS BREWING CO.
DUBOIS BUDWEISER & WURZBURGER
DUBOIS, PA., NEWARK, NJ., BUFFALO, NY.
KAUFMANN & STRAUSS
"AMERICAN MAID"
ROUND 13" PIE
PRE - PRO

DUBOIS BREWING CO.
DUBOIS BUDWEISER BEER
DUBOIS, PA.
MFG: ?
ROUND 12"
1930s

EAGLE BREWING COMPANY
OLD DUTCH PREMIUM BEER
CATASAUQUA, PA.
MFG: ?
ROUND 12"
1950s

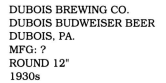

DUBOIS BREWING CO.
HAHNE'S PORTER
DUBOIS, PA.
CHAS. W. SHONK
ROUND 12"
PRE - PRO

EAGLE BREWERY
CATASAUQUA BEER
CATASAUQUA, PA.
ELECTRO - CHEMICAL ENGRAVING CO., INC.
ROUND 12"
MID - LATE 1930s

EAGLE BREWING COMPANY
OLD DUTCH BEER
CATASAUQUA, PA.
NOVELTY WORKERS UNION (AFOFL) COSHONCTON, OHIO
ROUND 12"
1950s

EMMERLING BREWING CO.
GROSSVADER OLD GERMAN LAGER BEER
JOHNSTOWN, PA.
KAUFMANN & STRAUSS
"DAS SCHMECKT GUT"
RECT. 11" x 14"
C. 1913

ELK BREWING CO.
RED BEER, ALE AND PORTER
KITTANING, PA (ST. MARYS)
AMERICAN ART WORKS
STOCK #77
ROUND 13" PIE
1909 - 1920

THE ERIE BREWING CO.
OLD DOBBIN ALE, KOEHLER'S ERIE BEER
ERIE, PA.
THE NOVELTY ADV. CO.
ROUND 14"
1930s

THE ELK RUN BREWING CO.
EXPORT BEER
PUNXUTAWNEY, PA.
CHAS. W. SHONK
ROUND 12" PIE
1902 - 1916

ESSLINGER'S, INC.
ESSLINGER REPEAL BEER
PHILADELPHIA, PA.
THE NOVELTY ADV. CO.
RECT. 11" x 14"
1933 - 1935

OTTO ERLANGER BREWING CO.
ERLANGER DELUXE BEER
PHILADELPHIA, PA.
CANCO
ROUND 13"
1940s

THE ERIE BREWING CO.
KOEHLER'S BEER, ALE, PORTER
ERIE, PA.
NOVELTY ADV. CO.
ROUND 14"
MID-1930s

THE ERIE BREWING CO.
KOEHLER'S BEER
ERIE, PA.
CANCO
ROUND 13"
1950s, 1960s

THE ERIE BREWING CO.
KOEHLER'S BEER
ERIE, PA.
CANCO
ROUND 12"
LATE 1950s, 1960s

ESSLINGER'S, INC.
ESSLINGER S PREMIUM BEER
PHILADELPHIA, PA.
BEACH COMPANY
ROUND 13"
1940s

FERNWOOD BREWING CO.
GOLDEN ALE BEER
FERNWOOD, PA. (now called LANSDOWN)
AMERICAN ART WORKS
ROUND 12"
1934 - 1941

FELL BREWING CO.
FELL BEER
CARBONDALE, PA. (now called SIMPSON)
ELECTRO - CHEMICAL ENGRAVING CO., INC.
ROUND 12"
LATE 1930s

ESSLINGER'S, INC.
ESSLINGER BEER
PHILADELPHIA, PA.
CCC
ROUND 12"
LATE 1950s, EARLY 1960s

FAIRVIEW BREWERY, AUGUST SCHNEIDER
AUGUSTINER BEER
READING, PA.
THE H. D. BEACH CO.
"BITTE NEHMAN SIE EINS"
ROUND 13 1/2"
1906 - 1912

FERNWOOD BREWING CO.
GOLDEN ALE BEER
FERNWOOD, PA.
ELECTRO - CHEMICAL ENGRAVING CO., INC.
ROUND 12"
1936

FINK BREWING COMPANY
FINK'S BEER & ALE
HARRISBURG, PA.
ELECTRO - CHEMICAL ENGRAVING CO., INC.
ROUND 12"
1933 - 1934

FORT PITT BREWING CO.
FORT PITT BEER
PITTSBURGH, PA.
P.O. 14 SHARPESBURG SUBURB
AMERICAN CAN CO.
ROUND 12"
1940s

FORT PITT BREWING CO.
OLD SHAY ALE
JEANNETTE, PA., SHARPESBURG, PA.
ELECTRO - CHEMICAL ENGRAVING CO., INC.
ROUND 13"
1937

FINK BREWING CO.
HARRISBURG, U.S.A.
AMERICAN ART WORKS
"THE CONNOISSEUR" ARTIST: COLES PORTER
OVAL 14" x 17"
C. 1913

FORT PITT BREWING CO.
OLD SHAY DELUXE BEER - ALE
JEANNETTE, PA.
A.C. CO.
ROUND 13"
LATE 1940s, EARLY 1950s

FRANKLIN BREWERY
FRANKLIN LAGER BEER
WILKES-BARRE, PA.
CHAS. W. SHONK
STOCK #63
ROUND 12"
1911 - 1919

FUHRMANN & SCHMIDT BREWING COMPANY, INC.
FREELAND BEER
FREELAND, PA.
ELECTRO - CHEMICAL ENGRAVING CO., INC.
ROUND 12"
1937

FRANKLIN BREWING CO., INC.
FRANKLIN BEER
WILKES-BARRE, PA.
MASON CAN CO.
ROUND 13"
1940s

FUHRMANN & SCHMIDT BREWING COMPANY, INC.
EAGLE RUN BEER, SPARKLING ALE & PORTER
SHAMOKIN, PA.
THE H. D. BEACH CO.
ROUND 14" PIE
1901 - 1906

FUHRMANN & SCHMIDT BREWING CO.
F & S BEER - ALE - PORTER
SHAMOKIN, PA.
ACME
ROUND 12"
EARLY 1930s

FUHRMANN & SCHMIDT BREWING CO.
F & S BEER - ALE
SHAMOKIN, PA.
UNIVERSAL TRAY AND SIGN CO.
ROUND 12"
1930s

FUHRMANN & SCHMIDT BREWING CO.
F & S BEER - ALE
SHAMOKIN, PA.
ELECTRO - CHEMICAL ENGRAVING CO., INC.
ROUND 12"
1937

FUHRMANN & SCHMIDT BREWING CO.
F & S BEER AND ALE
SHAMOKIN, PA.
UNIVERSAL TRAY AND SIGN CO.
ROUND 12"
LATE 1930s

FUHRMANN & SCHMIDT BREWING CO.
F & S BEER AND ALE
SHAMOKIN, PA.
UNIVERSAL TRAY AND SIGN CO.
ROUND 12"
1930s

WILLIAM GRETZ BREWING CO.
GRETZ BEER
PHILADELPHIA, PA.
BURDICK
ROUND 12"
LATE 1930s

JOS. H. GLENNON'S BREWERY
GLENNON'S BEER - ALE - PORTER
PITTSTON, PA.
THE MEEK COMPANY
"ST. VINCENT" STOCK #89
SQUARE 14" x 14"
C. 1908

WILLIAM GRETZ BREWING CO.
GRETZ BEER
PHILADELPHIA, PA.
MFG: ?
ROUND 13"
1940s

WILLIAM GRETZ BREWING CO.
GRETZ BEER
PHILADELPHIA, PA.
CCC
ROUND 12"
1950s

JOS. H. GLENNON'S BREWERY
GLENNON'S BEER
PITTSTON, PA.
AMERICAN ART WORKS
"THE BUFFALO HUNT"
SQUARE 14" x 14"
C. 1909

HAEFNER BREWING CO.
TIVOLI BEER
LANCASTER, PA.
MFG: ?
ROUND 13" PIE
LATE 1930s, EARLY 1940s

GOENNER & CO.
NEW LIFE BEER
JOHNSTOWN, PA.
KAUFMANN & STRAUSS
"THE NEW LIFE GIRL"
ROUND 12"
PRE - PRO

ROBERT H. GRAUPNER BREWERY
EXPORT BEER
HARRISBURG, PA.
THE MEEK CO. STOCK #42
OVAL 14" x 17"
1903 - 1905

HAZLE BREWING CO.
HAZLETON, PA.
CHAS. W. SHONK
ROUND 12" PIE
1900 - 1910

HEALTH BEVERAGE CO.
FAMOUS "READING BEER"
READING, PA
THE BURDICK CO.
"PURITY-HEALTH-VIGOR"
ROUND 12" PIE
1929 - 1933

JOHN HOHENADEL BREWERY, INC.
HOHENADEL BEER - ALE
PHILADELPHIA, PA.
A.C. CO.
ROUND 13"
1930s

JOHN HOHENADEL BREWERY INC.
HOHENADEL BEER, DOPPELBRAU,
INDIAN QUEEN ALE
PHILADELPHIA, PA.
ELECTRO - CHEMICAL ENGRAVING CO., INC.
ROUND 12"
1936

HOWELL AND KING CO.
JOYCE'S PERFECTION BEER
PITTSTON, PA.
N.Y. IMPORTING CO.
ROUND 12" PIE
1933 - 1935

HOWELL AND KING CO.
PERFECTION BEER
PITTSTON, PA.
MFG: ?
ROUND 13"
1933 - 1935

THE HOME BREWING CO.
BEER, ALE AND PORTER
SHENANDOAH, PA.
MFG: ?
ARTIST: CHAS. EHLEN
OVAL 14" x 17"
C. 1905

HORLACHER BREWING CO.
HORLACHER'S PURE BEER
ALLENTOWN, PA.
THE MEEK CO.
OVAL 14" x 17"
1902 - 1905

JACOB HORNUNG, TIOGA BREWERY
HORNUNG'S WHITE BOCK BEER
PHILADELPHIA, PA.
AMERICAN ART WORKS
STOCK #77
ROUND 13" PIE
1909 - 1920

JACOB HORNUNG BREWING CO.
HORNUNG'S BEER
PHILADELPHIA, PA.
A.C. CO.
ROUND 12"
1940s

JACOB HORNUNG BREWING CO.
HORNUNG'S BEER
PHILADELPHIA, PA.
MFG: ?
ROUND 12"
EARLY 1950s

IRON CITY BREWING CO.
RHEINGOLD BEER
LEBANON, PA.
THE H. D. BEACH CO.
ROUND 13" PIE
PRE - PRO

JACOB HORNUNG BREWING CO.
LONDONDERRY ALE
PHILADELPHIA, PA.
MFG: ?
ROUND 13"
LATE 1930s, EARLY 1940s

CHAS. D. KAIER CO., LTD.
KAIER'S BEER - ALE - PORTER
MAHANOY CITY, PA.
ACME
ROUND 12"
LATE 1930s, EARLY 1940s

CHAS. D. KAIER CO., LTD.
KAIER BEER, ALE AND PORTER
MAHANOY CITY, PA.
MFG: ?
STOCK #12
OVAL 14" x 17"
PRE - PRO

CHAS. D. KAIER CO., LTD.
BEER, ALE AND PORTER
MAHANOY CITY, PA.
MFG: ?
ROUND 12"
PRE - PRO

CHAS. D. KAIER CO.
KAIER'S BEER - ALE - PORTER
MAHANOY CITY, PA.
BURDICK CO., INC.
ROUND 12"
1940s

KAIER BREWING CO.
KAIER'S BEER
MAHANOY CITY, PA.
MFG: ?
ROUND 12"
1950s

JOHN KAZMAIER, BREWER
ALTOONA, PA.
MFG: ?
"A CLOSE GAME" STOCK #48
OVAL 14" x 17"
PRE - PRO

HERMAN KOSTENBADER & SONS, EAGLE BREWERY
KOSTENBADER EXPORT BEER
CATASAUQUA, PA.
THE MEEK CO.
STOCK #42
OVAL 14" x 17"
1902 - 1909

HERMAN KOSTENBADER & SONS, EAGLE BREWERY
STANDARD - EXPORT
CATASAUQUA, PA.
THE MEEK CO.
STOCK #77
SQUARE 14" x 14"
1902 - 1909

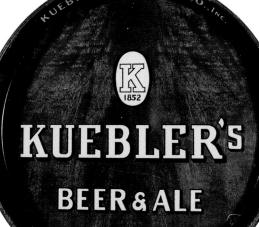

KUEBLER BREWING CO., INC.
KUEBLER BEER & ALE
EASTON, PA.
UNIVERSAL TRAY AND SIGN CO.
ROUND 12"
1930s

KUEBLER BREWING CO., INC.
KUEBLER BEER
EASTON, PA.
ELECTRO - CHEMICAL ENGRAVING CO., INC.
ROUND 12"
1939

LEBANON BREWING CO.
BEER, ALE AND PORTER
LEBANON, PA.
MFG: ?
ARTIST: CHAS. EHLEN
OVAL 14" x 17"
C. 1905

LEBANON VALLEY BREWING CO.
EXPORT BEER, ALE - PORTER
LEBANON, PA.
ELECTRO - CHEMICAL ENGRAVING CO., INC.
ROUND 13"
LATE 1930s, EARLY 1940s

LIBERTY BREWING CORP.
GLENNON'S BEER, ALE AND PORTER
PITTSTON, PA.
ROUND 13" PIE
1933 - 1934

KUEBLER BREWING CO., INC.
KUEBLER BEER
EASTON, PA.
CCC
ROUND 13"
LATE 1940s, EARLY 1950s

KUEBLER BREWING CO., INC.
KUEBLER BEER
EASTON, PA.
MFG: ?
ROUND 12"
EARLY 1940s

JOS. LIEBERMAN'S SONS
LIEBERMAN'S LAGER BEER
ALLENTOWN, PA
THE MEEK CO.
"GRISELDA"
SQUARE 14" x 14"
C. 1907

LACKAWANNA BEER AND ALE CORP.
LACKAWANNA BEER AND ALE
SCRANTON, PA.
UNIVERSAL TRAY AND SIGN CO.
ROUND 12"
1933 - 1943

LIEBERT & OBERT
MANAYUNK BEER AND ALE
PHILADELPHIA, PA.
UNIVERSAL TRAY AND SIGN
ROUND 12"
MID-LATE 1930s

LIEBERT & OBERT A.K.A.
COOPER BREWING CO.
COOPER'S FINE BEER
PHILADELPHIA, PA.
A.C. CO.
ROUND 12"
LATE 1930s, EARLY 1940s

THE LION INC. A.K.A. GIBBONS BREWERY
GIBBONS BEER AND ALE
WILKES-BARRE, PA.
CCC
ROUND 12"
1950s, 1960s

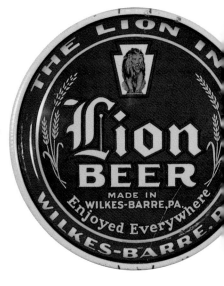

THE LION INC.
LION BEER
WILKES-BARRE, PA.
GENERAL BREWERS
SUPPLY CORP.

THE LION INC. A.K.A. GIBBONS BREWERY
GIBBONS BEER
WILKES-BARRE, PA.
CCC
ROUND 13"
LATE 1960s, 1970s

FRED C. LUCAS, CASTANEA BREWERY
CASTANEA BEER
LOCK HAVEN, PA.
MFG: ?
STOCK #6
OVAL 14" x 17"
PRE - PRO

THE LODER BREWING CO.
BEER, ALE, PORTER
COLUMBIA, PA.
CHAS. W. SHONK
"REMEMBER THE MAINE"
ROUND 12" PIE
1898 - 1901

MELLET & NICHTER BREWING CO.
BEER, ALE & PORTER
POTTSVILLE, PA.
AMERICAN ART WORKS
"THE INVITATION" STOCK #116
RECT. 11" x 14"
C. 1911

LUZERNE COUNTY BREWING CO.
BEER AND PORTER
WILKES-BARRE, PA.
MFG: ?
STOCK #5
OVAL 14" x 17"
1905 - 1910

MAUCH CHUNK BREWING CO.
MAUCH CHUNK BEER
E. MAUCH CHUNK, PA.
MFG: ?
ROUND 13" PIE
1934 - 1941

F. C. LUCAS BREWER, CASTANEA BREWERY
CASTANEA BEER
LOCK HAVEN, PA.
MFG: ?
STOCK #6
OVAL 14" x 17"
PRE - PRO

LUZERNE COUNTY BREWING CO.
EDELBRAU BEER
WILKES-BARRE, PA.
THE MEEK CO.
ROUND 13" PIE
1905 - 1909

MELLET-NICHTER BREWING CO.
BEER, ALE AND PORTER
POTTSVILLE, PA.
THE H. D. BEACH CO.
ROUND 12"
1908 - 1919

MOUNT CARMEL BREWERY
ANTHRACITE FINE BEER
MOUNT CARMEL, PA.
UNIVERSAL TRAY AND SIGN CO.
ROUND 12"
1940s

MOOSE BREWING CO.
PENNSY SELECT BEER
ROSCOE, PA.
IVAN B. NORDHAM COMPANY
"OUR MUTUAL FRIEND"
ROUND 12"
1903 - 1919

NEUSTADTL BREWING CORP.
GESUNDHEIT
STROUDSBERG, PA.
ELECTRO - CHEMICAL ENGRAVING CO., INC.
ROUND 12"
1933 - 1935

MOUNT CARBON BREWERY
BAVARIAN TYPE BEER
POTTSVILLE, PA.
THE BEACH CO.
ROUND 13"
EARLY - MID 1940s

NORTHAMPTON BREWING CO.
WURTZBURGER, PILSNER
NORTHAMPTON, PA.
CHAS. W. SHONK
RECT. 11" x 14"
PRE - PRO

MOUNT CARMEL BREWERY
METZGER'S OLD GERMAN
MOUNT CARMEL, PA.
UNIVERSAL TRAY AND SIGN CO.
ROUND 12"
1940s

NORTHAMPTON BREWERY CORP.
TRU-BLU BEER AND ALE
NORTHAMPTON, PA.
MFG: ?
ROUND 12"
1930s

LOUIS F. NEUWEILER'S SONS
NEUWEILER'S ALE - BEER - PORTER
ALLENTOWN, PA.
AMERICAN CAN CO.
ROUND 12" and 13" versions
1940s

NORTHAMPTON BREWING CO.
TRU-BLU BEER
NORTHAMPTON, PA.
AMERICAN ART WORKS
"AT YOUR SERVICE" STOCK #117
ROUND 13" PIE
C. 1911

NORTHAMPTON BREWERY CORP.
TRU-BLU BEER - ALE
NORTHAMPTON, PA.
CCC
BLUE BOY BY GAINSBOROUGH
ROUND 12"
1940s

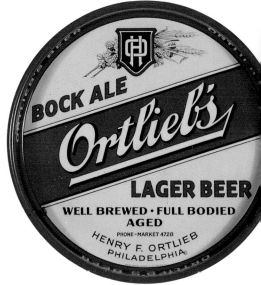

HENRY F. ORTLIEB BREWING CO.
ORTLIEB'S BOCK ALE, LAGER BEER
PHILADELPHIA, PA.
ELECTRO - CHEMICAL ENGRAVING CO., INC.
ROUND 13"
LATE 1930s, EARLY 1940s

THE OLD READING BREWERY, INC.
OLD READING BEER
READING, PA.
CCC
ROUND 12"
1950s

ORTLIEB BREWING CO.
ORTLIEB'S BOHEMIAN EXPORT BEER
MAUCH CHUNK, PA.
AMERICAN ART WORKS
"CARNATION GIRL"
ROUND 13" PIE
C. 1908

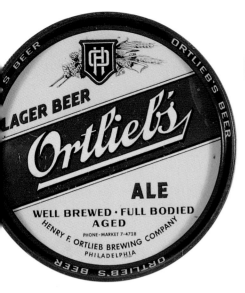

HENRY F. ORTLIEB BREWING CO.
ORTLIEB'S LAGER BEER - ALE
PHILADELPHIA, PA.
CCC
ROUND 12"
1940s

P. & H. BREWING CO.
P. & H. LEBANON BEER
LEBANON, PA.
MFG: ?
ROUND 13" PIE
1933 - 1934

PENNSYLVANIA CENTRAL BREWING CO.
PETER KRANTZ BREWERY
LAGER BEER & PORTER
CARBONDALE, PA. (now SIMPSON)
HAEUSERMANN LITHO
RECT. 11" x 14"
1905 - 1919

PHILADELPHIA BREWING CO.
MANZ OLD STOCK BEER & ALE
PHILADELPHIA, PA.
THE BURDICK CO.
ROUND 12"
LATE 1930s

PILSNER BREWING CO.
PILSNER-HEIM BEER AND PORTER
HAZELTON, PA.
THE MEEK CO.
"PRIDE OF THE FARM" STOCK #90
SQUARE 14" x 14"
C. 1908

PHILADELPHIA BREWING CO.
MANZ BEER - ALE
PHILADELPHIA, PA.
MFG: ?
ROUND 13"
1940s

PHILADELPHIA BREWING CO.
OLD STOCK BEER - ALE
PHILADELPHIA, PA.
THE BURDICK CO.
ROUND 12"
LATE 1930s

PILSNER BREWING CO.
HAZELTON PILSNER STYLE BEER
HAZELTON, PA.
ELECTRO - CHEMICAL ENGRAVING CO., INC.
ROUND 12"
LATE 1930s, EARLY 1940s

PITTSBURGH BREWING CO.
OLD KING COLE BEER
UNIONTOWN, PA.
AMERICAN CAN CO.
ROUND 12"
1935 - 1942

PITTSBURGH BREWING CO.
E & O BEER
PITTSBURGH, PA.
AMERICAN CAN CO.
ROUND 12"
1950s

PITTSBURGH BREWING CO.
TECH BEER
PITTSBURGH, PA.
MFG: ?
ROUND 12"
1960s

POTH BREWING CO., INC.
POTH'S BEER
PHILADELPHIA, PA.
UNIVERSAL TRAY AND SIGN CO.
ROUND 12"
1933 - 1936

PITTSTON BREWING CORPORATION
GLENNON'S BEER, OLD MULE ALE
PITTSTON, PA.
UNIVERSAL TRAY AND SIGN CO.
ROUND 12"
1934 - 1942

PITTSTON BREWING CORP.
GLENNON'S BEER, ALE AND PORTER
PITTSTON, PA.
H. D. BEACH CO.
RECT. 11" x 14"
1934 - 1942

PITTSBURGH BREWING CO.
IRON CITY BEER
PITTSBURGH, PA.
CCC
ROUND 12"
1960s

POTH BREWING CO., INC.
POTH ALL GRAIN BEER AND ALE
PHILADELPHIA, PA.
A.C. CO.
ROUND 12"
1933 - 1936

PURE SPRINGS BREWING CO., INC.
PURE SPRINGS BEER
FOUNTAIN SPRINGS, PA.
MFG: ?
RECT 11" x 14"
1933 - 1934

THE PUNXSUTAWNEY BREWING CO.
PUNXSUTAWNEY, PA.
AMERICAN ART WORKS
"THE INVITATION" STOCK #116
RECT. 11" x 14"
C. 1911

PREMIER BREWING COMPANY
PREMIER BEER
PHILADELPHIA, PA.
MFG: ?
RECT 11" x 14"
1912 - 1920

REICHARD & WEAVER BREWERY
WILKES-BARRE, PA.
AMERICAN ART WORKS
ROUND 13" PIE
C. 1911

REICHARD & WEAVER BREWERY
WILKES-BARRE, PA.
HAEUSERMANN LITHO
ROUND 13" PIE
PRE - PRO

RETTIG BREWING CO.
BEER, PORTER AND ALE
POTTSVILLE, PA.
MFG: ?
"OLD FRIENDS" STOCK #45
ROUND 13" PIE
PRE - PRO

RETTIG BREWING CO.
BEER, PORTER AND ALES
POTTSVILLE, PA.
MFG: ?
ROUND 12"
PRE - PRO

E. ROBINSON'S SONS
PILSNER BOTTLED BEER
SCRANTON, PA.
HAEUSERMANN LITHO
ROUND 13" PIE
PRE - PRO

E. ROBINSON'S SONS
PILSNER BOTTLED BEER
SCRANTON, PA.
HAEUSERMANN LITHO
ROUND 12"
1905 - 1919

THE SAYRE BREWING CO., LTD.
BEER, ALE AND PORTER
SAYRE, PA.
AMERICAN ART WORKS
"GOOD MORNING"
ROUND 13" PIE
C. 1913

ADAM SCHEIDT BREWING CO.
VALLEY FORGE BEER
NORRISTOWN, PA.
CANCO
ROUND 12"
1952 - 1954

ADAM SCHEIDT BREWING CO.
VALLEY FORGE BEER - RAMS HEAD ALE
NORRISTOWN, PA.
A.C. CO.
ROUND 13"
1940s

SCHILLER BREWING CO.
SCHILLER BEER
PHILADELPHIA, PA.
THE BURDICK CO.
ROUND 12"
1940 - 1941

ADAM SCHEIDT BREWING CO.
PRIOR LAGER BEER
NORRISTOWN, PA.
CANCO
ROUND 12"
1950 - 1955

C. SCHMIDT & SONS, INC.
SCHMIDT'S ALE
PHILADELPHIA, PA.
MFG: ?
ROUND 13"
1950s

C. SCHMIDT & SONS, INC.
SCHMIDT'S LIGHT BEER
PHILADELPHIA, PA.
CCC
ROUND 13"
1970s

SEITZ BREWING CO.
SEITZ BEER
EASTON, PA.
THE BURDICK CO.
ROUND 12" (also 13" PIE)
1934 - 1938

SEITZ BREWING CO.
SEITZ PALE BEER
EASTON, PA.
CHAS. W. SHONK
OVAL 13" x 16"
PRE - PRO

SEITZ BREWING CO.
YE OLD ALE, BOHEMIAN EXPORT LAGER,
SPECIAL BREW
EASTON, PA.
CHAS. W. SHONK
RECT. 11" x 14"
PRE - PRO

SOUTH BETHLEHEM BREWING CO.
SUPREME
BETHLEHEM, PA.
AMERICAN ART WORKS
STOCK #77
ROUND 13" PIE
1909 - 1920

SEITZ BREWING CO.
EASTON, PA.
THE MEEK CO.
"ST. VINCENT" STOCK #89
SQUARE 14" x 14"
1902 - 1909

SOUTH BETHLEHEM BREWING CO.
SUPREME
BETHLEHEM, PA.
MFG: ?
"THE HOME FAVORITE" STOCK #6
OVAL 14" x 17"
PRE - PRO

SOUTH BETHLEHEM BREWING CO.
SUPREME BEER
BETHLEHEM, PA.
CCC
ROUND 13"
1940s

SOUTH BETHLEHEM BREWING CO.
SUPREME BEER
BETHLEHEM, PA.
UNIVERSAL TRAY AND SIGN CO.
ROUND 12"
1940s

SOUTH FORK BREWING CO.
SOUTH FORK
SOUTH FORK, PA.
MFG: ?
RECT 11" x 14"
1933 - 1940

SOUVENIR - BEER DRIVERS UNION
LOCAL 132 - PHILADELPHIA VICINITY
PICTURE: CAPT. KONIG (inset)
"DEUTCHLAND"
MFG: ?
ROUND 13" PIE

SPRENGER BREWING CO.
RED ROSE BEER
LANCASTER, PA.
NOVELTY ADV. CO.
ROUND 14"
MID - LATE 1930s

STEGMAIER BREWING CO.
PORTER, STOCK LAGER, STANDARD LAGER
WILKES-BARRE, PA.
STANDARD ADV. CO.
(Factory-made hole in center of upper rim)
ROUND 13" PIE
1897 - 1901

STANDARD BREWING CO.
TRU-AGE BEER
SCRANTON, PA.
METALLOGRAPH CORP.
ROUND 12"
1930s

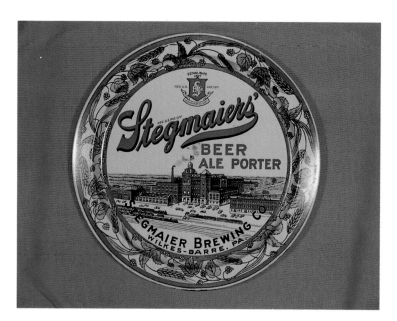

STEGMAIER BREWING CO.
STEGMAIERS BEER, ALE - PORTER
WILKES-BARRE, PA.
BURDICK
ROUND 13" PIE
1930s

STEGMAIER BREWING CO.
STEGMAIER GOLD MEDAL BEER
WILKES-BARRE, PA.
AMERICAN ART WORKS
ROUND 12"
1930s

STEGMAIER BREWING CO.
STEGMAIER GOLD MEDAL BEER
WILKES-BARRE, PA.
AMERICAN CAN CO.
ROUND 13" PIE
1933 - 1940

STEGMAIER BREWING CO.
STEGMAIER BEER
WILKES-BARRE, PA.
CHAS. W. SHONK
ROUND 12"
PRE - PRO

STEGMAIER BREWING CO.
STEGMAIER'S GOLD MEDAL BEER
WILKES-BARRE, PA.
MFG: ?
BAKELITE TRAY (IA)
ROUND 12"
1942 - 1945

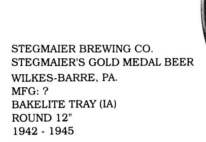

STEGMAIER BREWING CO.
STEGMAIER GOLD MEDAL BEER
WILKES-BARRE, PA.
CCC
ROUND 12"
1953

STEGMAIER BREWING CO.
STEGMAIER GOLD MEDAL BEER
WILKES-BARRE, PA.
CCC
ROUND 12"
1953

STEGMAIER BREWING CO.
STEGMAIER'S GOLD MEDAL BEER
WILKES-BARRE, PA.
OWENS - ILLINOIS CAN CO.
ROUND 13"
1940s

STEGMAIER BREWING CO.
STEGMAIER GOLD MEDAL BEER
WILKES-BARRE, PA.
CCC
ROUND 13"
1959

ST. MARY'S BEVERAGE CO.
ALES, LAGER, AULT DEUTCHER BEER
ST. MARY'S, PA.
H. D. BEACH CO.
ROUND 12"
1933 - 1940

ST. MARY'S BEVERAGE CO.
ST. MARYS BEER - ALE
ST. MARY'S, PA.
THE H. D. BEACH CO.
ROUND 12"
1933 - 1940

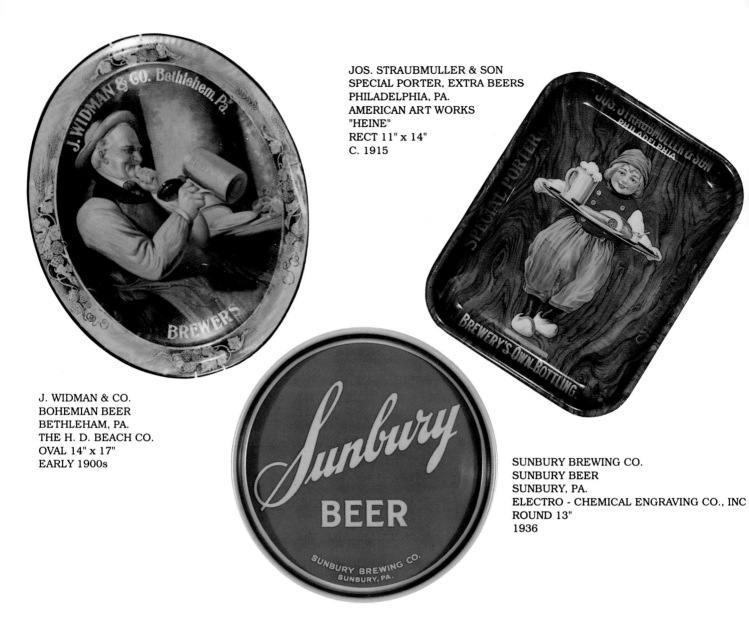

JOS. STRAUBMULLER & SON
SPECIAL PORTER, EXTRA BEERS
PHILADELPHIA, PA.
AMERICAN ART WORKS
"HEINE"
RECT 11" x 14"
C. 1915

J. WIDMAN & CO.
BOHEMIAN BEER
BETHLEHAM, PA.
THE H. D. BEACH CO.
OVAL 14" x 17"
EARLY 1900s

SUNBURY BREWING CO.
SUNBURY BEER
SUNBURY, PA.
ELECTRO - CHEMICAL ENGRAVING CO., INC
ROUND 13"
1936

STRAUB BREWERY, INC.
STRAUB BEER
ST. MARYS, PA.
CANCO
ROUND 13"
1950s

TREMONT BREWING COMPANY
FRANK SCHAUB'S LAGER BEER
TREMONT, PA.
CHAS. W. SHONK
ROUND 12" PIE
1899 - 1908

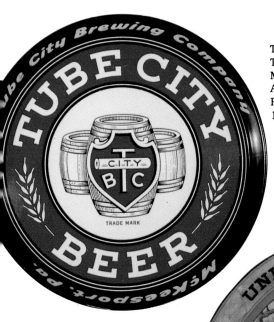

TUBE CITY BREWING COMPANY
TUBE CITY BEER
McKEESPORT, PA.
A.C. CO.
ROUND 12"
1933 - 1935

UHL'S BREWERY (UHL'S ESTATE)
UHL'S VIENNA SPECIAL
BETHLEHEM, PA.
THE MEEK CO.
"A GOOD JUDGE"
SQUARE 14" x 14"
C. 1908

UNION BREWERY, HOWELL & KING CO.
PITTSTON, PA.
MFG: ?
STOCK #36
OVAL 14" x 17"
PRE - PRO

VIKING BREWERY, INC.
VIKING BEER
CATASAUQUA, PA.
NOVELTY ADVERTISING CO.
RECT 11" x 14"
1933 - 1935

WACKER BREWING CO.
WACKER ALL MALT LAGER BEER
LANCASTER, PA.
AMERICAN ART WORKS
ROUND 12"
1940s

WAYNE BREWING CO.
BEER - ALE - PORTER
ERIE, PA.
CHAS. W. SHONK
"MAD ANTHONY WAYNE BLOCKHOUSE, ERIE, PA."
ROUND 12"
1907 - 1919

WEISBROD & HESS BREWING CO.
WEISBROD BEER
PHILADELPHIA, PA.
ELECTRO - CHEMICAL ENGRAVING CO., INC.
ROUND 12"
1933 - 1938

WEISBROD & HESS BREWING CO.
PHILADELPHIA, PA (ATLANTIC CITY, WILDWOOD N.J. BRANCHES)
AMERICAN ART WORKS
"JOIN ME"
ROUND 13" PIE
C. 1909

D. G. YUENGLING & SON, INC.
YUENGLING'S BEER, ALE AND PORTER
POTTSVILLE, PA.
A.C. CO.
ROUND 12"
1941

D. G. YUENGLING & SON
YUENGLING'S BEER, ALE AND PORTER
POTTSVILLE, PA.
AMERICAN ART WORKS
"BEAUTY AND THE BEAST"
ROUND 13" PIE
C. 1911

D. G. YUENGLING & SON, INC.
YUENGLING'S BEER, ALE AND PORTER
POTTSVILLE, PA.
AMERICAN CAN CO.
ROUND 12"
1940s

D. G. YUENGLING & SON
YUENGLING'S BEER, ALE AND PORTER
POTTSVILLE, PA.
AMERICAN ART WORKS
"BEAUTY AND THE BEAST"
ROUND 13" PIE
C. 1911

J. WIDMAN & CO.
BOHEMIAN BEER
BETHLEHEM, PA.
MFG: ?
ROUND 12" PIE
PRE - PRO

J. WIDMAN & CO.
BOHEMIAN BEER
BETHLEHEM, PA.
MFG: ?
ROUND 12" PIE
PRE - PRO

D. G. YUENGLING & SON, INC.
YUENGLING'S PRIZE BEER
POTTSVILLE, PA.
AMERICAN CAN CO.
ROUND 12"
1940s

D. G. YUENGLING & SON, INC.
YUENGLING'S PREMIUM BEER
POTTSVILLE, PA.
CCC
ROUND 13"
MID - 1940s

Bibliography

Anderson, Sonja and Will. *Beers, Breweries and Breweriana.* Carmel, NY. 1969

Anderson, Will. *Berr New England.* Portland, ME, 1988.

Anderson, Will. *The Beer Book.* Princeton, NJ: Pyne Press, 1973.

Bull, Donald, Freidrich, Manfred, Gottschalk, Robert. *American Breweries.* Trumbull, CT: Bullworks, 1984.

Burden, Paul. *The Beer Tray Book.* Medfield, MA: Norfield Publishing, 1992.

Congdon-Martin, Douglas. *America for Sale.* West Chester, PA: Schiffer Publishing LTD, 1991.

Cope, Jim. *Collectible Old Advertising.* Orange, TX, 1973.

House of Collectibles, Editors. *The Official Price Guide to Beer Cans & Collectibles, Fourth Edition.* New York: Random House, 1986.

Julia, James D. Inc. *Important Advertising Auction to Include "The Beer Museum".* Fairfield, ME, 1992.

MacRae's, Blue Book. Chicago, IL, Various years.

Mortimeyer, W.R. *Advertising "Worth" Collecting.* Cuba, Ms. 1972.

Muzio, Jack. *Collectible Tin Advertising Trays.* Santa Rosa, CA, 1972.

Ojala, Reino. *20 Years of American Beers - The '30s & '40s.* Minneapolis, MN: Published by Author, Winslow Printing.

Polansky, Tom. *Advertising Trays.* Loma Linda, CA, 1971.

Robertson, James D. *The Connoisseur's Guide to Beer.* Aurora, IL: Caroline House Publisher, INC., 1982.

Robertson, James D. *The Great American Beer Book.* New York, NY: Warner Books, 1978.

Thomas Register of American Manufacturers. Various years.

Weiner, Michael A. *The Taster's Guide to Beer.* New York, NY: Collier Books, 1977.

Yenne, Bill. *Beers of North America.* New York, NY: Gallery Books, 1986.

Periodicals

American Breweriana Journal - Bimonthly publication of the American Breweriana Association, Inc.

Beer Can Collectors News Report - Bimonthly publication of the Beer Can Collectors of America.

The Keg - Quarterly publication of the Eastern Coast Breweriana Association.

The Breweriana Collector - Quarterly publication of the National Association of Breweriana Advertising.

Beer Cans Monthly - Maverick Publishing Co., Bulkner, MO, Dabbs, Robert L., Editor (No longer in publication).

Brewery Collectibles Magazine - Class Publishing Co., Colmer, PA. Cameron, Jeffery C., Editor. Published bimonthly. (No longer in publication).

* Multiple volumes of all the above, totaling around 300, were reviewed for tray photos.

Value Guide

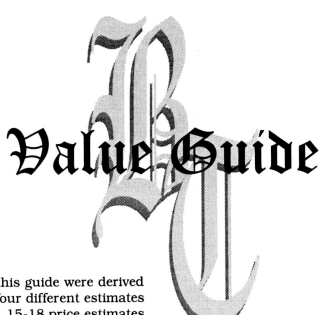

The value figures in this guide were derived by averaging a minimum of four different estimates for each tray. In many cases, 15-18 price estimates were averaged. Opinions varied considerably, and in one case, ranged from $400 to $2200. Prices are for trays in pristine condition and values drop drastically for examples in less than perfect condition.

New Jersey

page 25
- T: $325
- C: $10
- B: 600

page 26
- TL: $5
- TR: $10
- CL: $40
- CR: $45
- B: $45

page 27
- TL: $40
- TR: $85
- CL: $125
- CR: $75
- B: $250

page 28
- TL: $25
- TR: $100
- C: $350
- BL: $110
- BR: $60

page 29
- TL: $200
- TR: $275
- C: $110
- BL: $80
- BR: $65

page 30
- TL: $65
- TR: $35
- CL: $35
- CR: $40
- B: $45

page 31
- TL: $20
- TR: $120
- C: $25
- BL: $350
- BR: $25

page 32
- TL: $10
- TR: $10
- C: $20
- BL: $15
- BR: $15

page 33
- TL: $10
- TR: $120
- C: $120
- BL: $60
- BR: $125

page 34
- TL: $80
- TR: $150
- C: $150
- BL: $300
- BR: $550

page 35
- TL: $70
- TR: $500
- C: $80
- BL: $150
- BR: $150

Massachusetts

page 36
- TL: $450
- TR: $85
- BL: $600
- BR: $75

page 37
- TL: $200
- TR: $300
- C: $45
- BL: $300
- BR: $275

page 38
- TL: $100
- TR: $50
- C: $175
- BL: $35
- BR: $75

page 39
- TL: $50
- TR: $40
- C: $200
- BL: $60
- BR: $80

page 40
- TL: $75
- TR: $55
- C: $180
- BR: $40

page 41
- TL: $40
- C: $45
- BL: $55
- BR: $175

page 42
- TL: $50
- TR: $45
- C: $180
- BL: $225
- BR: $80

page 43
- TL: $60
- TR: $50
- CL: $35
- CR: $40
- B: $35

page 44
- TL: $40
- TR: $45
- C: $20
- BL: $80
- BR: $40

page 45
- TL: $500
- TR: $110
- C: $160
- BL: $215
- BR: $400

page 46
- TL: $150
- TR: $225
- C: $185
- BL: $150
- BR: $65

page 47
- TL: $450
- TR: $100
- C: $200
- BL: $450
- BR: $65

page 48
- TL: $600
- TR: $350
- C: $70
- BL: $75
- BR: $85

Rhode Island

page 49
- TL: $1500
- TR: $275
- BL: $200
- BL: $150

page 50
- TL: $200
- TR: $180
- C: $275
- BL: $100
- BR: $35

page 51
- TL: $35
- TR: $175
- CL: $300
- CR: $325
- B: $30

page 52
- TL: $300
- TR: $25
- C: $250
- BL: $40
- BR: $40

page 53
- TL: $35
- TR: $375
- C: $35
- BL: $40
- BR: $350

page 54
- TL: $600
- TR: $275
- C: $350
- BL: $450
- BR: $150

page 55
- TL: $900
- TR: $700
- BL: $750
- BR: $150

New York

page 56
- TL: $60
- TR: $260
- C: $40
- BL: $249
- BR: $85

page 57
- TL: $225
- TR: $200
- CL: $300
- CR: $60
- BR: $45

page 58
- TL: $40
- TR: $35
- C: $40
- BL: $50
- BR: $50

page 59
- TL: $125
- TR: $150
- CL: $65
- CR: $100
- B: $290

page 60
- TL: $165
- TR: $625
- C: $60
- BL: $250
- BR: $35

page 61
- TL: $300
- TR: $450
- C: $350
- BL: $100
- BR: $250

page 62
- TL: $30
- TR: $450
- C: $250
- BL: $350
- BR: $125

page 63
- TL: $400
- TR: $300
- C: $250
- BL: $275
- BR: $60

page 64
- TL: $45
- TR: $125
- CL: $65
- CR: $125
- B: $30

page 65
- TL: $140
- TR: $40
- CL: $65
- CR: $200
- B: $225

page 66
- TL: $100
- TR: $225
- C: $75
- BL: $75
- BR: $90

page 67
- TL: $30
- TR: $55
- C: $200
- BL: $35
- BR: $35

page 68
- TL: $30
- C: $20
- BL: $15
- BR: $15

page 69
- TL: $15
- TR: $10
- C: $10
- BL: $230
- BR: $5

page 70
- TL: $275
- TR: $70
- CL: $140
- CR: $150
- BL: $125
- BR: $80

page 71
- TL: $125
- TR: $150
- C: $150
- BL: $250
- BR: $125

page 72
- TL: $150
- TR: $65
- C: $70
- BL: $40
- BR: $50

page 73
- TL: $125
- TR: $25
- C: $30
- BL: $280
- BR: $180

page 74
TL: $50
TR: $250
C: $250
BL: $40
BR: $50

page 75
TL: $50
TR: $375
CL: $225
CR: $650
BL: $475
BR: $500

page 76
TL: $30
TR: $30
C: $40
BL: $20
BR: $500

page 77
TL: $225
TR: $125
C: $70
BL: $50
BR: $175

page 78
TL: $175
TR: $50
C: $80
BL: $30
BR: $100

page 79
TL,TR,CR,BR: $5
BL: $90

page 80
TL: $90
TR: $350
C: $130
BL: $85
BR: $80

page 81
TL: $100
TR: $375
C: $275
CR: $100
B: $65

page 82
TL: $125
TR: $600
C: $150
BL: $200
BR: $100

page 83
TL: $150
TR: $200
C: $550
BL: $150
BR: $150

page 84
TL: $40
TR: $50
BL: $375
BR: $35

page 85
TL,TR: $15
C: $150

B: $200

page 86
TL: $35
CL: $15
CR: $15
BL, BR: $20

page 87
TL: $100
TR: $350
C: $275
BL: $275
BR: $375

page 88
TL: $575
TR: $20
C: $25
BL: $25
BR: $5

page 89
TL: $15
TR: $50
CL: $35
BL: $500

page 90
TL: $350
TR: $10
C: $350
BL: $10
BR: $15

page 91
TL: $10
TR: $25
C: $250
BL: $200
BR: $5

page 92
TL: $25
TR: $30
C: $5
BL: $60
BR: $45

page 93
TL: $250
TR: $700
BL: $25
BR: $15

page 94
TL: $75
TR: $35
C: $45
BL: $45
TR: $40

page 95
TL: $30
TR: $60
C: $85
BL: $40
BR: $20

page 96
TL: $650
TR: $300
C: $45
BL: $250
BR: $15

page 97

T: $300
C: $80
BL: $100
BR: $175

New Hampshire

page 98
T: $85
BL: $60
BR: $70

page 99
T: $450
BL: $45
BR: $350

Connecticut

page 100
TL: $800
TR: $700
C: $800
BL: $125

page 101
TL: $125
TR: $275
C: $250

page 102
TL: $175
TR: $175
C: $210
BL: $125
BR: $125

page 103
TL: $425
TR: $250
C: $350
BL: $275
BR: $225

page 104
T: $325
CL: $125
CR: $85
BL: $125
BR: $65

page 105
TL: $150
TR: $175
C: $135
B: $20

page 106
TL: $600
TR: $250
C: $180
BL: $250
BR: $425

page 107
TL: $175
TR: $160
C: $175
BL: $150
BR: $200

page 108
T: $235
C: $150
B: $100

page 109

TR: $125
CL: $110
CR: $170
B: $125

Pennsylvania

page 110
TL: $400
TR: $450
BL: $450
BR: $450

page 111
T: $150
CL: $550
CR: $65
B: $275

page 112
T: $500
CL: $600
CR: $575
B: $150

page 113
TL: $120
TR: $125
CL: $120
CR: $350
B: $45

page 114
T: $500
CL: $85
CR: $125
BL: $150
BR: $75

page 115
TL: $400
TR: $60
CL: $60
BL: $75
BR: $75

page 116
TL: $40
TR: $50
C: $50
BL: $225
BR: $100

page 117
TL: $125
TR: $175
CR: $50
BL: $250
BR: $150

page 118
TL: $50
TR: $550
C: $160
BL: $450
BR: $135

page 119
T: $125
CL: $50
CR: $20
BL: $125
BR: $10

page 120
TL: $35
TR: $100

C: $20
BL: $650
BR: $200

page 121
TL: $60
TR: $225
CL: $75
CR: $60
B: $50

page 122
TL: $40
TR: $275
C: $150
BL: $400
BR: $60

page 123
TL: $140
TR: $115
CL: $75
CR: $110
B: $75

page 124
TL: $40
TR: $550
C: $35
B: $35

page 125
TL: $600
CL: $350
CR: $425
B: $400

page 126
TL: $85
TR: $45
CL: $75
CR: $120
B: $150

page 127
C: $700

page 128
C: $128

page 129
TL: $325
TR: $45
CL: $65
CR: $450
B: $90

page 130
TL: $20
TC: $600
TR: $350
CL: $40
CR: $20
B: $700

page 131
C: $550

page 132
TL: $350
TC: $60
TR: $65
BL: $300
BC: $100
BR: $125

page 133

TL: $120
TC: $40
TR: $200
C: $80
BL: $500
BR: $35

page 134
TL: $10
TR: $140
C: $10
BL: $900
BR: $500

page 135
T: $475
CL: $625
CR: $140
BL: $650
BR: $300

page 136
TR: $425
TL: $130
C: $300
BL: $35
BR: $200

page 137
TL: $120
TR: $725
BL: $35
BC: $575
BR: $50

page 138
TL: $70
TR: $35
C: $40
B: $350

page 139
TL: $20
TC: $200
TR: $300
CL: $65
CR: $350
B: $50

page 140
TL: $55
TR: $60
BL: $100
BC: $110
BR: $15

page 141
TR: $25
CL: $120
CR: $400
BL: $10
BR: $125

page 142
TL: $300
TR: $140
CL: $250
CR: $575
B: $550

page 143
C: $600

page 144
T: $300
B: $350

page 145

C: $300
page 146
TR: $450
TL: $20
C: $30
BL: $35
BR: $125

page 147
TL: $10
TC: $10
TR: $150
BL: $650
BR: $550

page 148
TL: $300
TR: $210
C: $350
BL: $35
BR: $35

page 149
T: $160
B: $130

page 150
TL: $130
TR: $1,500
C: $70
BL: $150
BR: $65

page 151
TL: $60
TC: $425
TL: $45
BL: $15
BR: $15

page 152
T: $20
C: $25
B: $85

page 153
C: $100

page 154
TL: $250
TR: $250
C: $200
BL: $125
BR: $500

page 155
TL: $160
TR: $475
C: $750
BL: $350
BR: $225

page 156
TL: $75
TC: $150
TR: $425
BL: $500
BC: $85
BL: $100

page 157
TL: $400
TR: $450
BL: $60
BR: $15